ENERGY SCIENCE, ENGINEERING AND TECHNOLOGY

DIESEL EMISSIONS REDUCTION EFFORTS

ENERGY SCIENCE, ENGINEERING AND TECHNOLOGY

Additional books in this series can be found on Nova's website under the Series tab.

Additional E-books in this series can be found on Nova's website under the E-book tab.

ENVIRONMENTAL REMEDIATION TECHNOLOGIES, REGULATIONS AND SAFETY

Additional books in this series can be found on Nova's website under the Series tab.

Additional E-books in this series can be found on Nova's website under the E-book tab.

ENERGY SCIENCE, ENGINEERING AND TECHNOLOGY

DIESEL EMISSIONS

REDUCTION EFFORTS

JARRETT F. PRICE
AND
SHERYL E. MITCHELL
EDITORS

Nova Science Publishers, Inc.
New York

For permission to use material from this book please contact us:
Telephone 631-231-7269; Fax 631-231-8175
Web Site: http://www.novapublishers.com

Library of Congress Cataloging-in-Publication Data

ISBN: 978-1-62081-561-8

Published by Nova Science Publishers, Inc. † New York

CONTENTS

PREFACE

Reducing emissions from diesel engines is one of the most important public health challenges facing the country. Despite the EPA's stringent diesel engine and fuel standards taking effect over the next decade, the 20 million engines already in use will continue to emit large amounts of nitrogen oxide and particulate matter, both of which will contribute to serious public health problems for years to come. Fortunately, a variety of cost-effective technologies can dramatically reduce harmful emissions, save fuel, and help our nation meet its clean air and sustainability goals. This book analyzes the cost-effectiveness of various diesel emission reduction strategies, wherein the EPA compares the amount of federal grant funds used, to the lifetime tons reduced, to evaluate the effectiveness of federal investments in spurring diesel emission reductions.

Chapter 1 - Reducing emissions from diesel engines is one of the most important public health challenges facing the country. Despite EPA's stringent diesel engine and fuel standards taking effect over the next decade, the 20 million engines already in use will continue to emit large amounts of nitrogen oxides (NO_X) and particulate matter (PM)—both of which will contribute to serious public health problems for years to come.

Fortunately, a variety of cost-effective technologies can dramatically reduce harmful emissions, save fuel, and help our nation meet its clean air and sustainability goals. To meet these challenges, the U.S. Environmental Protection Agency (EPA) established the National Clean Diesel Campaign (NCDC). NCDC consists of both regulatory programs to address new engines and innovative nonregulatory programs to address the millions of diesel engines already in use. EPA standards apply to *new* diesel engines, and because these engines can last a long time, solutions are needed to reduce harmful emissions from the existing fleet. These innovative approaches promote a variety of emission reduction strategies such as retrofitting, repairing, replacing, and repowering engines; reducing idling; and switching to cleaner fuels.

Through a dynamic network of Regional Collaboratives, whose development EPA initiated, environmental groups, industry, and government were inspired and motivated—despite their sometimes conflicting perspectives—to unite behind a common goal. NCDC mobilized diverse and unusual partners with historic differences to work together, creating broad support based on the urgency of the public health problem and bringing new technologies into use years earlier than would otherwise have occurred.

In 2008, Congress appropriated funding for the Diesel Emissions Reduction Act (DERA) Program under the Energy Policy Act of 2005 (EPAct 2005) to reduce emissions from heavy-duty diesel engines in the existing fleet. In the first year of the DERA program, EPA awarded $49.2 million for diesel emission reduction programs across the country. EPA gave priority to projects that:

- Demonstrated a clear public health benefit in areas with high population density and poor air quality.
- Fostered cost-effective strategies that maximized the useful life of a certified engine configuration, verified technology, or emerging technology.
- Conserved diesel fuel.
- Used cleaner fuels.

The fiscal year (FY) 2008 funding enabled EPA to award 119 grants, which will lead to emission reductions of approximately 46,000 tons of NO_X and 2,200 tons of fine particulate matter ($PM_{2.5}$). The PM-related emission reductions translate into a significant quantifiable public health benefit of approximately $580 million to $1.4 billion. Additionally, projects funded under these grants will save 3.2 million gallons of fuel per year for a cost savings to operators of more than $8 million each year (at $2.50 per gallon). More than 14,000 diesel-powered vehicles and pieces of equipment are cleaner as a result of the first year of this program. More than two-thirds of these vehicles are trucks and school buses that were upgraded through EPA's sector outreach programs such as Clean School Bus USA.

Chapter 2 - Exhaust from diesel engines is a harmful form of air pollution. EPA has issued emissions standards for new diesel engines and vehicles, but older mobile sources of diesel emissions—such as trucks and buses—continue to emit harmful pollution. Programs at DOE, DOT, and EPA provide funding for activities that reduce diesel emissions, such as retrofitting existing diesel engines and vehicles. The existence of these programs at multiple agencies has raised questions about the potential for unnecessary duplication. In response to a mandate in the Diesel Emissions Reduction Act of 2010, GAO examined the (1) extent of duplication, overlap, fragmentation, or gaps, if any, among federal grant, rebate, and loan programs that address mobile source diesel emissions; (2) effectiveness of federal funding for activities that reduce mobile source diesel emissions; and (3) extent of collaboration among agencies that fund these activities. GAO analyzed program data, documents, and relevant laws and regulations and interviewed agency officials. GAO also reviewed three diesel-related tax expenditures.

Chapter 3 - The Safe, Accountable, Flexible, Efficient Transportation Equity Act: A Legacy for Users (SAFETEA-LU) was enacted August 10, 2005, as Public Law 109-59. SAFETEA-LU creates new opportunities as well as responsibili-ties for States, Metropolitan Planning Organizations (MPOs), and other agencies involved in the selection of projects funded through the Congestion Mitigation and Air Quality Improvement Program (CMAQ). Among the changes, Section 1808 of SAFETEA-LU, for the first time, makes CMAQ funding available for many nonroad diesel retrofits for nonroad vehicles, engines, and construction equipment that are used on Title 23 projects in certain nonattainment areas. While acknowledging State and MPO roles and authority for making final project selections, Section 1808 also addresses funding priority. "(A) IN GENERAL – States and metropolitan plan ing organizations shall give priority in distributing funds received for congestion mitigation and air quality projects and programs from apportionments derived from applications of sections 104(b)(2)(B) and 104(b)(2)(C) to – (i) diesel retrofits, particularly where necessary to facilitate contract compliance, and other cost-effective emission reduction activities, taking into consideration air quality and health effects; and (ii) cost-effective congestion mitigation activities that provide air quality benefits. (B) SAVINGS – This paragraph is not intended to disturb the existing authorities and roles of governmental agencies in making final project selections."

In: Diesel Emissions: Reduction Efforts
Editors: J. F. Price and S. E. Mitchell
ISBN: 978-1-62081-561-8

Chapter 1

HIGHLIGHTS OF THE DIESEL EMISSIONS REDUCTION PROGRAM*

National Clean Diesel Campaign

EXECUTIVE SUMMARY

National Clean Diesel Campaign

Reducing emissions from diesel engines is one of the most important public health challenges facing the country. Despite EPA's stringent diesel engine and fuel standards taking effect over the next decade, the 20 million engines already in use will continue to emit large amounts of nitrogen oxides (NO_X) and particulate matter (PM)—both of which will contribute to serious public health problems for years to come.

Fortunately, a variety of cost-effective technologies can dramatically reduce harmful emissions, save fuel, and help our nation meet its clean air and sustainability goals. To meet these challenges, the U.S. Environmental Protection Agency (EPA) established the National Clean Diesel Campaign (NCDC). NCDC consists of both regulatory programs to address new engines and innovative nonregulatory programs to address the millions of diesel engines already in use. EPA standards apply to *new* diesel engines, and

* This is an edited, reformatted and augmented version of the Report to Congress, Energy Policy Act of 2005, Title VII, Subtitle G, Sections 791–797, United States Environmental Protection Agency, EPA-420-R-09-006, dated August 2009.

because these engines can last a long time, solutions are needed to reduce harmful emissions from the existing fleet. These innovative approaches promote a variety of emission reduction strategies such as retrofitting, repairing, replacing, and repowering engines; reducing idling; and switching to cleaner fuels.

Through a dynamic network of Regional Collaboratives, whose development EPA initiated, environmental groups, industry, and government were inspired and motivated—despite their sometimes conflicting perspectives—to unite behind a common goal. NCDC mobilized diverse and unusual partners with historic differences to work together, creating broad support based on the urgency of the public health problem and bringing new technologies into use years earlier than would otherwise have occurred.

In 2008, Congress appropriated funding for the Diesel Emissions Reduction Act (DERA) Program under the Energy Policy Act of 2005 (EPAct 2005) to reduce emissions from heavy-duty diesel engines in the existing fleet. In the first year of the DERA program, EPA awarded $49.2 million for diesel emission reduction programs across the country.[1] EPA gave priority to projects that:

- Demonstrated a clear public health benefit in areas with high population density and poor air quality.
- Fostered cost-effective strategies that maximized the useful life of a certified engine configuration, verified technology, or emerging technology.
- Conserved diesel fuel.
- Used cleaner fuels.

The fiscal year (FY) 2008 funding enabled EPA to award 119 grants, which will lead to emission reductions of approximately 46,000 tons of NO_X and 2,200 tons of fine particulate matter ($PM_{2.5}$). The PM-related emission reductions translate into a significant quantifiable public health benefit of approximately $580 million to $1.4 billion. Additionally, projects funded under these grants will save 3.2 million gallons of fuel per year for a cost savings to operators of more than $8 million each year (at $2.50 per gallon). More than 14,000 diesel-powered vehicles and pieces of equipment are cleaner as a result of the first year of this program. More than two-thirds of these vehicles are trucks and school buses that were upgraded through EPA's sector outreach programs such as Clean School Bus USA.

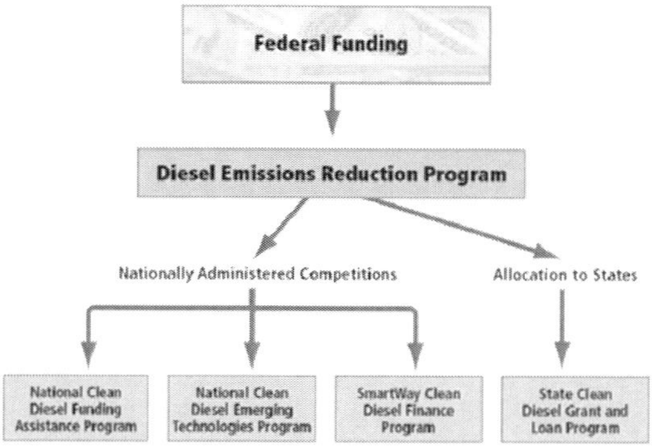

National Clean Diesel Campaign Funding Structure.

DERA directs EPA to fund two different components: a national competition and a state allocation program. The national program, with 70 percent of the funding, consists of three separate competitions:

1. The National Clean Diesel Funding Assistance Program
2. The National Clean Diesel Emerging Technologies Program
3. The SmartWay Clean Diesel Finance Program

The State Clean Diesel Grant and Loan Program utilizes the remaining 30 percent of the funding.

NCDC programs deploy proven technologies much earlier than would occur otherwise, accelerate the use of emerging technologies, and provide innovative financial incentives that make the business case for doing the right thing for the environment. States are key partners in the success of clean diesel programs and are given the opportunity through the state program to enhance their leadership on air quality issues.

National Clean Diesel Programs

National Clean Diesel Funding Assistance Program

The response to EPA's National Clean Diesel Funding Assistance Program was overwhelmingly positive. This program allows communities to meet their most pressing needs to reduce harmful diesel emissions. Communities benefit by activities as diverse as retrofitting school buses so that children's exposures are reduced, repowering locomotives used at seaports to save fuel as well as reduce emissions in neighborhoods, and replacing high-emitting construction equipment used to build hospitals and our nation's roads.

EPA's seven Regional Collaboratives held grant competitions. In FY 2008, applicants nationwide submitted 236 applications, requesting more than $144 million and offering approximately $81 million in matching funds. Demand for funding under this component of DERA exceeded available funds by 5:1. The variety of needs and approaches reflects the diversity of air quality challenges facing local communities across the country. After careful evaluation and ranking, EPA awarded approximately $30 million to 60 applicants, who provided matching funds of approximately $35 million. The high level of response to this program shows the desire communities have to clean up their fleets. The response also shows the ability of federal dollars to attract funds from other partners to accelerate the introduction of cost-effective clean diesel technologies.

National Clean Diesel Emerging Technologies Program

The emerging technology program, for which EPA may make available up to 10 percent of the funding, fosters the deployment of cutting-edge technologies and encourages private-sector investment in innovation. The program promotes the deployment of innovative technologies (those not yet verified or certified by EPA or the California Air Resources Board) by providing funding to develop and evaluate these technologies in the field. EPA received 10 eligible applications that requested $5 million and offered more than $1.1 million in matching funds. EPA ultimately awarded six grants totaling $3.7 million with $1.1 million provided as matching funds. These grants will demonstrate new technologies, including technologies that reduce NO_X. This program will help expand the currently limited retrofit options for nonroad engines, such as those used in construction equipment and on marine vessels.

SmartWay Clean Diesel Finance Program

For the first time, DERA gives EPA the authority to establish national low-cost revolving loan and other financing programs to provide funding to fleets to re-duce diesel emissions. EPA exercised this authority by awarding grants where the recipients could provide innovative financing methods for clean diesel technologies. For the first time, partners are able to provide financial incentives (e.g., low-cost revolving loans, rebates, below-market rates) that stretch the federal dollar further. A small investment of federal funds leverages significant private investment and spurs the purchase of fuel-saving and emission reduction technologies for trucks. Under the SmartWay Clean Diesel Finance Program competition, EPA received four grant applications requesting approximately $9.5 million that would leverage $44 million in outside funds, to establish innovative financing programs. EPA awarded three grants totaling approximately $3.4 million, leveraging $19 million. These grants establish national financing programs that provide funding to small- and medium-sized trucking companies to purchase clean diesel technologies. These trucking companies include many small businesses that would otherwise be unlikely to update their fleets with cleaner, more fuel-efficient equipment.

State Clean Diesel Grant and Loan Program

Recognizing that many states, such as California, Texas, Washington, Illinois, Maine, and New York, have developed successful clean diesel programs over the years, Congress included a state-only component in DERA to recognize the vital role played by states. In contrast to the national competitive program, the State Clean Diesel Grant and Loan Program is a formula allocation program. By meeting certain requirements, any state is eligible to participate. According to the statute, if adequate appropriations are available, 30 percent of any funds appropriated for DERA must be allocated to the states through this program.

In the first year of the state program, all 50 states elected to participate and established new clean diesel projects. The programs responded to state needs, focusing on a variety of sectors, such as school buses, construction, freight, refuse haulers, and transit buses. Based on funding availability and the states' ability to provide matching funds, individual state grant amounts ranged from

$196,880 to $492,200. EPA funded the state programs with almost $13 million. Thirty-three states matched the federal dollars by providing more than $6.3 million in matching funds altogether.

Cost-Effective Results of Proven Technological Solutions

In this report, EPA analyzes the overall cost-effectiveness of various diesel emission reduction strategies. EPA compares the amount of federal grant funds used to the lifetime tons reduced to evaluate the effectiveness of federal investments in spurring diesel emission reductions. The Agency concludes that the innovative finance provisions are an important new approach with an impressive ability to leverage significant additional resources.

In the first year of funding, clean diesel strategies have proven to be about as cost-effective in reducing air pollution and health impacts associated with diesel emissions as other EPA programs, as shown in the table on page 7. Under the National Clean Diesel Funding Assistance Program, EPA calculated the average cost-effectiveness of federal investment to be approximately $27,700 per ton of PM reduced over the lifetime of the project and just under $2,000 per lifetime ton of NO_X. In the SmartWay Clean Diesel Finance Program, the cost-effectiveness was about $9,000 in federal funds per lifetime ton of PM and $400 in federal funds per lifetime ton of NO_X. The average cost of federal investment for the state program was approximately $16,700 per lifetime ton of PM reduced and $600 per lifetime ton of NO_X. These per ton costs compare very favorably with strategies used to attain national ambient air quality standards, such as stationary source standards that range from $1,000 to $20,000 and as high as $100,000 per ton of $PM_{2.5}$ on an annualized basis.[2, 3] However, DERA projects are on average[4] less cost-effective for PM than regulatory programs designed to set emissions standards for new diesel engines, such as the 2007 Heavy-Duty Highway diesel emissions standards. EPA has estimated that this regulation has a cost-effectiveness of $14,200 per ton PM and $2,100 per ton NO_x.

The first year of implementation for the Diesel Emissions Reduction Program has been extremely successful. By harnessing overwhelming support from environmental organizations, state and local governments, industry, technology vendors, and other groups, EPA has been able to magnify the available dollars and strategically fund important emission reduction and fuel-saving projects. In total, EPA awarded 119 grants in the first year of the DERA program. More than 14,000 vehicles and pieces of nonroad equipment in a wide array of sectors will be cleaner as a direct result of this program. New clean diesel programs have been established in every state, and technologies have been advanced for the future. In addition, the American Recovery and Reinvestment Act of 2009 (Recovery Act) provided $300 million in new funding for the Diesel Emissions Reduction Program. EPA has already awarded more than $85 million through the State Clean Diesel Grant program, and those funds are currently at work creating green jobs and reducing air pollution. EPA is in the process of awarding national competitive grants under the Diesel Emissions Reduction Program with additional goals of creating and preserving jobs and promoting economic recovery. In response to this competitive program, EPA received more than 600 applications requesting approximately $2 billion and offering more than $2 billion in matching funds. This clearly demonstrates the high level of interest in clean diesel programs across the country. Implementation of the EPAct 2005's Diesel Emissions Reduction Program is a critical step toward reaching our national clean air goals and protecting public health and the environment for all Americans for generations to come.

Average Cost-Effectiveness of DERA Projects

DERA Program	**$PM_{2.5}$**	**NO_X**
National Clean Diesel Funding Assistance Program	$27,700	$2,000
SmartWay Clean Diesel Finance Program	$9,000	$400
State Clean Diesel Grant and Loan Program	$16,700	$600
Regulatory Programs	**$PM_{2.5}$**	**NO_X**
Heavy-Duty Engine and Highway Diesel Fuel Requirements	$14,200	$2,100
PM NAAQS RIA	$3,000-15,000	–

Note 1: Values reported for the DERA program are for federal funds per ton pollutant reduced over project lifetime. Lifetime tons are not discounted. This report presents two different types of cost-effectiveness: federal funds per lifetime tons and total cost per ton. The comparisons for the regulatory programs are in annualized costs per ton.

Note 2: Values were rounded to nearest $100. Federal funds do not include matching amounts.

Note 3: The Emerging Technologies program is designed to demonstrate new technologies. At this phase in the development of these technologies, costs would be expected to be higher than when at a more fully mature commercial scale of deployment stage. Therefore, the cost comparisons with the other national program components have been omitted.

1. INTRODUCTION: REDUCING DIESEL EMISSIONS GARNERS BROAD SUPPORT

From the farm to the interstate highway to the neighborhood grocery store, we find diesel engines in every corner of society. Diesel engines power the movement of goods across the nation and help construct the buildings in which we live and work. Diesel engines aid in building the roads on which we travel and power the buses that carry millions of children to school safely each day. While diesel engines provide the mobility and power Americans require, are more fuel-efficient than gasoline engines, and serve as the workhorse of the nation's economy, exhaust from these engines sullies our skies with pollutants that harm our health and damage the environment.

Because of our reliance on diesel engines, reducing emissions from them is one of the most important public health challenges facing the country. Despite the U.S. Environmental Protection Agency's (EPA's) stringent diesel engine and fuel standards, which, for new engines beginning with engine model year 2007, are being phased in over the next decade, 20 million engines already in use continue to emit relatively large amounts of nitrogen oxides (NO_X) and fine particulate matter ($PM_{2.5}$). NO_X, which contributes to both ozone and particulate formation, as well as directly emitted $PM_{2.5}$, can lead to serious health conditions such as triggering asthma and worsening heart and lung disease. In addition, diesel engines emit black carbon, which might contribute to global climate change.

The problem of diesel emissions is not limited to a few discrete geographic areas or to one segment of the population. Harmful diesel emissions contribute to poor air quality in much of the country. No fewer than 141 million Americans—nearly half the population of the United States— live in areas that are designated as nonattainment with the eight-hour ozone standard and/or the $PM_{2.5}$ standard, based on the National Ambient Air Quality Standards (NAAQS).[5]

Due to the sheer numbers of engines in use and the volume of pollutants they emit, the health effects and environmental pollution stemming from diesel emissions could be substantial. Nation-wide, in 2009 diesel emissions from mobile sources alone will account for approximately 300,000 tons of directly emitted $PM_{2.5}$ and 6.4 million tons of NO_X, which contribute to the formation of ozone and additional fine particles. These emissions will come from approximately 20 million engines operating in 2009, including approximately 13 million on-highway vehicles, 7 million nonroad engines, and 47,000 locomotive and marine engines. (See Figures 1 and 2.) Reducing diesel emissions quickly is vital to helping communities reach their public health goals.

Fortunately, a variety of cost-effective technologies can dramatically reduce harmful emissions, save fuel, and help our nation meet its clean air and sustainability goals. In 2000, to address the concerns of both new and existing diesel engines, EPA created the National Clean Diesel Campaign (NCDC), a partnership program that incorporates traditional regulatory approaches and innovative nonregulatory approaches to achieve results. The regulatory aspect of the program requires new engines and their fuels to meet stringent technology-based standards to reduce the amount of emissions released.

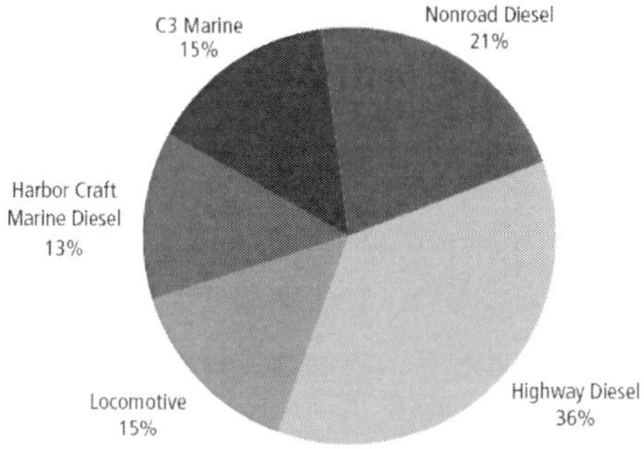

Figure 1. 2009 NOX Mobile Source Diesel Emissions for 50 States (6,400,000 tons).

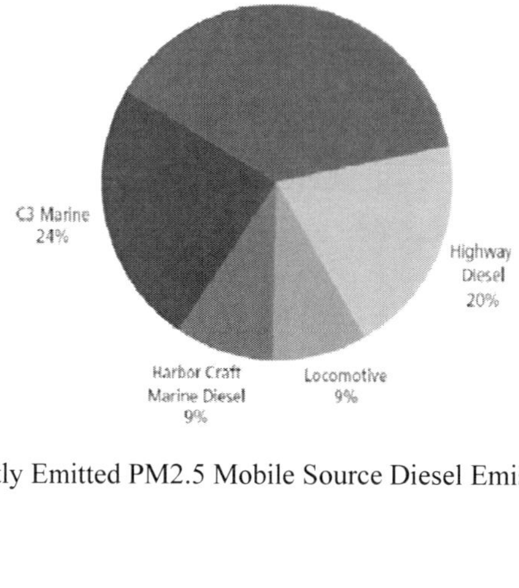

Figure 2. 2009 Directly Emitted PM2.5 Mobile Source Diesel Emissions for 50 States (300,000 tons).

The innovative component of the program promotes the use of a variety of techniques to reduce emissions, including retrofitting, repairing, replacing, and repowering vehicles and equipment; reducing idling; and switching to cleaner fuels. To encourage these actions, NCDC cultivates the involvement of national, state, and local partners in the public and private sectors.

In 2008, for the first time ever, Congress appropriated funding under the Energy Policy Act of 2005 (EPAct 2005) to reduce emissions from diesel engines in the nation's existing fleet. In the first year of the program, the EPA's NCDC distributed $49.2 million to initiate diesel emission reduction projects and programs across the country. Reflecting the goals in the statute, EPA gave priority to projects that:

- Demonstrate a clear public health benefit and apply to areas with high population density and poor air equality.
- Use cost-effective strategies that maximize the useful life of a certified engine configuration, verified technology, or emerging technology.
- Conserve diesel fuel.
- Use cleaner fuels.

The EPAct of 2005, Title VII, Subtitle G, Section 794, rquires that the EPA Administrator submit a report to Congress evaluating the implementation of the Diesel Emissions Reduction Act (DERA) Program. This report fulfills that requirement.

In the inaugural year of the program, EPA awarded 119 grants that are already achieving significant emission reductions, as detailed in Appendices A through F. As part of this effort, all 50 states have begun new state clean diesel program grants. The suite of national and state diesel emission reduction programs is a vital part of state and local efforts to reduce harmful air pollution. These first DERA grants are expected to significantly reduce emissions in communities across the country.

As shown in Table 1, the total emission reductions are noteworthy. The FY 2008 funding for the Diesel Emissions Reduction Program is estimated to reduce approximately 46,000 tons of NO_X and 2,200 tons of $PM_{2.5}$ over the life of the program. These emission reductions translate into a significant quantifiable public health benefit of approximately $580 million to $1.4 billion in quantifiable PM-related health benefits over the life of the program.[6]

Diesel Exhaust Health Effects

Direct emissions from diesel engines, especially $PM_{2.5}$, NO_X, and sulfur oxides (SO_X), contribute to health problems. In addition, NO_X contributes to the formation of ozone and PM through chemical reactions.

$PM_{2.5}$ has been associated with an increased risk of pre-mature mortality, increased hospital admissions for heart and lung disease, and increased respiratory symptoms. Long-term exposure to components of diesel exhaust, including diesel PM and diesel exhaust organic gases, are likely to pose a lung cancer hazard. Exposure to ozone can aggravate asthma and other respiratory diseases, leading to more asthma attacks, the use of additional medication, more severe symptoms that require a doctor's attention, more lost school and work days, more visits to the emergency room, increased hospitalizations, and even premature mortality. People in many areas of the United States experience short-term (one to three hours) and prolonged ozone exposures (six to eight hours), which have been linked to diminished lung function, greater respiratory symptoms, and increased hospital visits. Repeated exposure to ozone can increase susceptibility to respiratory infection and lung inflammation and can aggravate preexisting asthma. At sufficient concentrations, ozone can even cause permanent damage to the lungs, including the development of chronic respiratory illnesses. Children, outdoor workers, those who exercise outdoors, people with heart and lung disease, and the elderly are most at risk.

In addition, the FY 2008 grants will save more than 3.2 million gallons of fuel per year, which equates to a savings in fuel costs to operators of more than $8 million per year (at $2.50 per gallon). This corresponds to 35,600 tons of carbon dioxide (CO_2) per year reduced. These fuel savings calculations only reflect the savings from idle reduction technologies and do not reflect the improved fuel economy that may result from engine replacements; thus the actual fuel savings might be higher than shown.

The map in Figure 3 shows the location of the projects for the first year of the program. These projects are helping communities lower their public health risks from air pollution, accelerating the use of proven technologies, making engines cleaner and more efficient, encouraging new cutting-edge technologies, and utilizing new financial incentives to reach more fleets.

Table 1. Estimated Air Emission Reductions of FY 2008 NCDC Programs over Lifetime of Projects (Lifetime Tons)

Program	NO_X	PM	HC*	CO**	CO_2†
National Clean Diesel Funding Assistance Program	14,800	1,100	1,900	7,300	113,600
National Clean Diesel Emerging Technologies Program	600	30	30	100	—
SmartWay Clean Diesel Finance Program	9,600	400	500	1,600	238,200
State Clean Diesel Grant and Loan Program	21,000	800	2,000	5,700	112,900
Total	**46,000**	**2,200**	**4,400**	**14,700**	**464,600**

Note: Values were rounded to the nearest 100 lifetime tons or to one significant figure. Sums might not add due to rounding. Lifetime tons are not discounted.

* Hydrocarbon

** Carbon monoxide

† Does not include CO_2 reductions from engine replacements, repowers, or vehicle replacements.

Figure 3. FY 2008 DERA Projects Across the Country.

2. EPA'S NATIONAL CLEAN DIESEL CAMPAIGN: A FIRST-OF-ITS-KIND CAMPAIGN LIVES UP TO ITS GOALS

Overview

EPA'S NCDC responds to the national need to reduce harmful diesel emissions. In April 2000, EPA began its Voluntary Diesel Retrofit Program with a few targeted projects. Using a small sample of the existing national fleet of school buses, delivery trucks, and transit buses, EPA demonstrated emission reduction technologies and cleaner fuels likely to be used for all new engines seven years before the 2007 effective date for the new engine standards.

Through this initial collaboration on retrofit projects, partners realized that they had found a way to put aside their sometimes-conflicting perspectives to unite behind a common goal. Fueled by this initial success, NCDC mobilized diverse partners with a wide range of perspectives—such as engine manufacturers, environmental groups, emission technology vendors, fuel suppliers, private fleet owners, state and local governments, and transportation officials—to work together, creating awareness of the urgency of the public health problem and accelerating the use of technologies years earlier than otherwise would have occurred. EPA provided seed money and technical expertise to create markets for new technologies and bring stakeholders together around an issue of mutual interest.

This tiny spark caught flame. Since 2000, EPA has awarded approximately 300 demonstration grants under the Clean Air Act for diverse sectors such as transit and school buses and marine, construction, and freight vehicles and equipment. These projects reduced unnecessary idling and promoted the use of alternative and cleaner fuels, retrofitting engines and replacing old engines or equipment with new, cleaner versions. Each year additional partners have joined NCDC, adding and showcasing new technologies. EPA tailored incentives to key sectors such as school buses, marine ports, construction, freight, transit, and agriculture. These successful demonstration projects resulted in the use of ultra-low-sulfur fuel in parts of 20 states well before the EPA mandate took effect. This cleaner diesel fuel directly reduces emissions from engines and enables the use of the most effective emission control devices. Now the use of this fuel is widespread and fleets are experiencing maintenance and emissions benefits of cleaner fuels.

Clean School Bus USA Is Key to a Healthier Ride for Kids

Due to the innovation and success of EPA's Clean School Bus USA, begun in 2003 as the flagship program of the National Clean Diesel Campaign, approximately 3 million children ride cleaner school buses today.

Clean School Bus USA's goal is to significantly reduce children's exposure to harmful exhaust from diesel school buses. By establishing partnerships with local and state governments, school transportation officials, fuel providers, and equipment and engine manufacturers, EPA created a highly successful, incentive-based program to reduce the pollution from the nation's 400,000 oldest and dirtiest school buses and to protect the 24 million children who ride buses daily.

As part of the program, EPA awarded 157 grants throughout six years, totaling $31 million, to demonstrate pollution control techniques. EPA has distributed its Key to a Healthier Ride educational materials to more than 400 school districts nationwide, reaching more than 74,000 school bus drivers and affecting more than 83,000 diesel school buses. These school districts are also formally recognizing bus drivers who successfully reduce school bus idling. For more information, visit: www.epa.gov/otaq/school bus/index.htm.

Verifying That Emission Reduction Technologies Work

Testing and evaluating the efficacy of diesel emission reduction technologies is a cornerstone of NCDC. From the beginning, fleet owners demanded to know that the technologies they wanted or were expected to use would work and would not harm their operations. State and local governments wanted assurance that they could rely on these strategies now and in the future to meet their air quality goals. In response to these needs, EPA created the

Retrofit Technology Verification Program in 2002 to evaluate objectively the effectiveness of emission control technologies.

Through this program, EPA assures users that the actual emission benefits from retrofit technologies match those claimed by the manufacturer. The Agency also evaluates technologies that assist fleets in reducing idling. In addition, the California Air Resources Board (CARB) provides its own list of verified emission control technologies (www.arb.ca.gov/diesel/verdev/verdev.htm), which EPA recognizes. By verifying technologies, leading successful demonstrations, and operating recognition and education programs, these agencies have helped develop markets for new, greener technologies.

Yet this was only the beginning. Congress recognized the important role the verification process plays by making it a foundation of the implementation and integrity of the Diesel Emissions Reduction Program. Thus, in the national grant program, EPA *requires* the use of technologies that have been verified or certified by EPA or CARB when federal funds are applied. Congress also wanted to continue to spur innovation and encouraged the support and development of emerging technologies under DERA.

Clean Ports USA Helps Marine Ports Save Money and Decrease Emissions

Due to EPA's Clean Ports USA program (part of the National Clean Diesel Campaign), leading marine ports are saving fuel, increasing operational efficiency, and decreasing diesel emissions. In addition, several prominent ports have developed clean air action plans, which include greenhouse gas reduction goals, to assist them in reducing air pollution in gateway communities.

CLEANPORTSUSA

In January 2005, EPA partnered with the American Association of Port Authorities to create Clean Ports USA as an outreach- and incentive-based program designed to reduce emissions from existing diesel engines operated at marine ports. EPA is working with port authorities; terminal operators; and shipping, trucking and rail companies to promote environmentally superior technologies through education, incentives, and financial assistance. For more information, visit: www.epa.gov/cleandiesel/ports.

SmartWay Transport Partnership

EPA developed the SmartWay Transport Partnership Program to address the environmental impact of the freight system in the United States. The Smart-Way program is a voluntary public–private initiative designed to improve the environmental performance of the freight delivery system through money-saving, market-based approaches. The goal of the program is to reduce emissions by promoting cost-effective strategies that reduce fuel consumption and air pollution, using such methods as eliminating unnecessary idling, installing emission control devices, and improving freight logistics.

The SmartWay program works hand in hand with the National Clean Diesel Campaign to promote diesel emission reduction strategies. The SmartWay program has pioneered innovative approaches to emission reduction, such as the creation of SmartWay-approved trucks to identify trucks with superior environmental performance and the development of low-cost financing for the purchase of fuel saving and emission control devices.

The innovative financial strategies offer increased access to lenders who can provide financing for SmartWay-verified technologies and provide truck owners with low-cost financing for the purchase of SmartWay Upgrade Kits (which include idle reduction technologies, advanced aerodynamic technologies, and low-rolling resistance tires) or the purchase of cleaner used diesel trucks with aftertreatment devices. The SmartWay program also recognizes partners who achieve superior environmental performance. For more information, visit: www.epa.gov/smartway.

In establishing its verification program, EPA realized that verifying a technology only prior to its use is not enough. The technologies need "real world" testing to ensure that they maintain their performance throughout their useful life.

As retrofit technologies are introduced into the market in higher volume, verifying the field performance and long-term durability of these products through in-use testing is increasingly essential. EPA requires all manufacturers to rigorously test each verified technology at two different stages during its useful life to confirm the percent of emissions reduction and evaluate durability. In-use testing offers confidence that verified retrofit technologies have been proven in the real world and will maintain that performance throughout their useful lives.

As a sign of the strength of the initial tests and of the technologies themselves, results from in-use testing are consistent with the original verified levels of emission reductions. In addition, EPA has been able to increase the level of reduction listed on its Web site for some of the technologies, given their outstanding performance in the in-use program.

To date, EPA and CARB have verified more than 50 emission control technologies. In addition, over the past seven years, EPA's SmartWay Transport Partnership has evaluated idle reduction technologies through grants, engineering analysis, and peer-reviewed reports. As of March 2009, EPA has designated six categories of idle reduction technologies as eligible for grant funding: 1) electrified parking spaces (truck stop electrification); 2) shore side power for ships, also known as cold ironing or Alternative Maritime Power; 3) auxiliary power units and generator sets; 4) fuel- operated heaters; 5) battery air-conditioning systems; and 6) thermal storage systems.

Tapping Into Regional Collaboratives

With the foundation of cost-effective, reliable technologies, EPA's partners wanted to reach more fleets in more sectors to reduce harmful diesel emissions. As a next step, beginning in 2003, EPA's regions began organizing Regional Clean Diesel Collaboratives with states, local nonprofit organizations, private industry, and municipalities. (See Figure 4 for a map.) By tying into a network of regional stakeholders, this collaborative structure is well-suited for achieving significant emission reductions across large geographic areas. Members of these collaboratives have agreed to collectively leverage additional funds and take a local approach to diesel emission mitigation. These collaboratives play a vital role in EPA's proactive, incentive-based approach to achieving superior environmental results. Information on the collaboratives can be found in Appendix F.

Figure 4. Regional Clean Diesel Collaboratives.

Regulating Diesel Engines and Fuels

While cost-effective technology and strong coalitions of committed stakeholders form the back-bone of NCDC, another important component is a series of regulations designed to reduce pollution emitted from new diesel engines and their fuels.

Under the Clean Air Act, EPA developed and implements a suite of stringent regulations to mitigate emissions from new diesel engines and their fuels. These regulations apply to a variety of new engines, including those in trucks, buses, construction equipment, locomotives, harbor craft, and large ships. Over the long term, these standards will yield enormous public health and environmental benefits.

When fully implemented, these regulatory programs will reduce NO_X by about 7 million tons per year, $PM_{2.5}$ by more than 300,000 tons, and SO_X by about 800,000 tons. By 2030, when the regulatory programs are fully phased in and implemented, the net public health benefits total approximately $186 billion per year, including annually preventing 26,000 premature deaths, approximately 20,000 hospitalizations, and more than 3.3 million days lost from work due to respiratory problems as described in EPA's Regulatory Impact Assessments.

In general, these rules require stringent standards for PM and NO_X, based on levels achievable with advanced diesel aftertreatment technologies, such as diesel particulate filters (DPFs) for PM and lean NO_X traps or selective catalytic reduction (SCR) for NO_X. Additionally, each of these rules ultimately requires reducing fuel sulfur levels to 15 parts per million (ppm) to enable the use of advanced aftertreatment technologies.

NCDC's goal is to bring these impressive benefits to the American people earlier by accelerating the adoption of proven technologies through its strategic regulatory approach.

3. The Energy Policy Act of 2005, Diesel Emissions Reduction Act Program: Congressional Support Buoys Programs Nationwide

Background

The EPAct of 2005 provides EPA new grant and loan authority to promote diesel emission reductions and authorizes appropriations of up to $200 million per year to the Agency under the Diesel Emissions Reduction Program provisions (for FY 2007 through FY 2011). Congress appropriated funds under this statute for the first time in FY 2008, in the amount of $49.2 million. This portion of the report highlights key components of the statute and explains how EPA is implementing the program with state partners pursuant to Congress's direction in the statute.

DERA includes a national program and a state allocation program. The national program includes 70 percent of appropriated funds, and the state program makes available 30 percent. Under this act, EPA can offer competitive grants and, for the first time, low-cost revolving loans to eligible organizations and entities on a competitive basis.

To spur innovation, DERA allows for up to 10 percent of the national funds to be spent on "emerging technologies." Additionally, not less than 50 percent of the funds available for the national program must be used for the benefit of public fleets. DERA requires that engine configurations and technologies supported by the national program be verified or certified by EPA or CARB.

The Energy Policy Act of 2005

Under the EPAct 2005, EPA gives priority to those diesel emission reduction programs that:

1. Maximize public health benefits.
2. Are the most cost-effective.
3. Are in areas with high population density and poor air quality (including nonattainment areas or areas that require maintenance of national ambient air quality standards for a criteria pollutant, federal Class I areas, or areas with toxic air pollutant concerns).
4. Are in areas that receive a disproportionate quantity of air pollution from diesel fleets, including truck stops, ports, rail yards, terminals, and distribution centers, or that use a community-based multi-stakeholder collaborative process to reduce toxic emissions.
5. Include a certified engine configuration or verified technology that has a long expected useful life.
6. Maximize the useful life of any certified engine configuration or verified technology used or funded by the eligible entity.
7. Conserve diesel fuel.
8. Utilize ultra-low sulfur diesel fuel (15 ppm of sulfur content) ahead of EPA's mandate (for nonroad or marine projects).

Relevant sections of the EPAct 2005 that govern the DERA program are found in Title VII, Subtitle G.

Section 791, "Definitions"
Section 792, "National Grant and Loan Programs"
Section 793, "State Grant and Loan Program"
Section 794, "Evaluation and Report"
Section 795, "Outreach and Incentives"
Section 796, "Relationship to Clean Air Act"
Section 797, "Authorization for Appropriated Funds (2007–2011)"

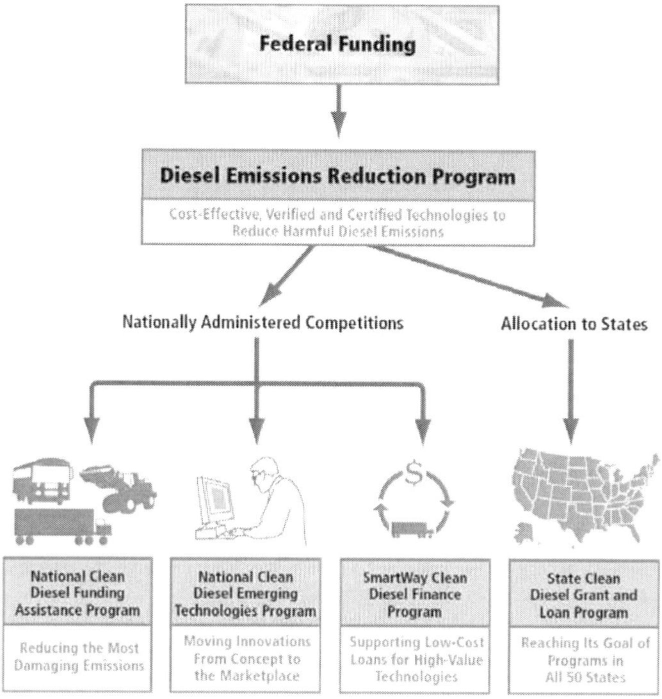

Figure 5. National Clean Diesel Campaign Funding Structure.

While the national program offers competitive grants in three categories, the state program is a single allocation program, which means that base funding is distributed to states using a specific formula based on participation, and incentive funding is also available for any states that match their base funding. Funds not claimed under this program—either if states decline to participate and/or do not match the base funding—revert to the national program. By meeting certain requirements, all of the 50 states and the District of Columbia are eligible to participate.[7]

Under the statute, EPA provides guidance to the states about the process for applications, permissible use of funds, and the cost-effectiveness of various emission reduction technologies. EPA established an annual deadline for submitting applications, an approval process, and a streamlined renewal process. Discretion is given to the governors for apportioning funds between grants and loans in each state. EPA published a *Federal Register* notice (73 FR 12728, March 10, 2008) that discussed how it intended to implement the state programs.

With the foundations of its successful Clean Air Act demonstration programs, verification program, sector-based focus, Regional Collaboratives, and SmartWay brand, NCDC was well-poised to deploy cost-effective diesel emission reduction strategies even more broadly under DERA.

EPA's goal for the first year of DERA funding was to establish diesel emission reduction programs in every state and accelerate the adoption of clean diesel technologies in the existing fleet. The EPAct 2005 statute establishes a general framework for the Diesel Emissions Reduction Program, charging EPA to develop other details for how the total program is implemented. Ultimately, EPA designed four programs:

1. The National Clean Diesel Funding Assistance Program
2. The National Clean Diesel Emerging Technologies Program
3. The SmartWay Clean Diesel Finance Program
4. The State Clean Diesel Grant and Loan Program

Emissions Reduced from Nonroad Cargo Handling Equipment at Pacific Northwest Ports

Through its Clean Ports USA Program, EPA awarded the Puget Sound Clean Air Agency an $850,000 grant to fund replacements and retrofits, such as diesel particulate filters, of nonroad truck engines for cargo-handling equipment at Puget Sound ports.

This grant helps the port authorities achieve the goals in their Northwest Ports Clean Air Plan. Additionally, the Ports of Seattle and Tacoma and the Puget Sound Clean Air Agency are providing matching funds totaling $318,000.

CLEANPORTSUSA

These programs deploy proven technologies much earlier than would otherwise occur to meet critical local air quality needs. They accelerate the use of emerging technologies and provide innovative financial incentives that make the business case for reducing diesel emissions. Furthermore, states are key partners in the success of clean diesel programs, and the state program enhances their leadership in reducing harmful diesel emissions.

The National Clean Diesel Funding Assistance Program: Reducing the Most Damaging Emissions

EPA received an overwhelmingly positive response to its Requests for Proposals (RFPs) under the National Clean Diesel Funding Assistance Program. This program offered the majority of the funding available, through grant competitions handled by the Agency's seven Regional Collaboratives. The Agency received 236 applications nationwide, requesting more than $144 million and offering $81 million in matching funds. Thus, applicants requested $5 for every $1 available. Applicants requested funding to address their most pressing needs, such as retrofitting school buses so that children's exposures are reduced on their way to school, repowering locomotives used at seaports to save fuel and reduce emissions in neighborhoods while still delivering freight used in every part of commerce, and replacing high-emitting construction equipment used to build hospitals and our nation's roads.

Table 2. FY 2008 Awards for the National Clean Diesel Funding Assistance Program by EPA Region

Region	Collaborative	Number of Applications	Number of Awarded Grants	Dollars Awarded
1	Northeast	16	6	$1,970,484
2		33	9	$3,268,402
3	Mid-Atlantic	24	6	$3,277,166
4	Southeast	39	7	$3,789,289
5	Midwest	38	13	$4,879,049
6	Blue Skyways	20	3	$2,719,880
7		10	3	$1,986,802
8	Rocky Mountain	11	5	$1,755,645
9	West Coast	30	5	$3,953,079
10		15	3	$1,440,007
Total		**236**	**60**	**$29,039,803**

Illinois School Buses Reap the Benefits of DERA Grants

EPA's Clean School Bus USA grants to Illinois EPA's Green Fleets Program, including school buses, totaled $678,600, plus a $654,700 match from Illinois EPA, funding a variety of clean diesel technologies, including diesel oxidation catalysts, diesel particulate filters, closed crankcase ventilation, and auxiliary power units.

The Illinois Green Fleets Program formally recognizes businesses, governmental entities, and organizations that use clean, domestic, renewable fuel vehicles in their fleet. The program also recognizes Illinois vehicle fleets equipped with diesel retrofits.

Public Fleets Directly Benefit from DERA National Program Funds

Public fleets across the nation—including transit buses, school buses, refuse haulers, municipal trucks, snowplows, and fire trucks—are emitting less pollution in towns and cities as a direct result of the Diesel Emissions Reduction Program.

In FY 2008, the Agency awarded 74 percent of competitively awarded national funds to public fleets. Section 792 (b)(2) of the EPAct 2005 requires that no less than 50 percent of the funds for the national program be spent for the benefit of public fleets.

Many partners stepped forward to deploy clean diesel technologies in public fleets to lead by example.

Making the public-sector vehicles that do the nation's work the cleanest they can be creates a powerful example for other fleets to imitate.

After screening for eligibility and carefully ranking the applications, EPA selected 60 recipients across the country in the fall and winter of 2008—within nine months of the program start date—and awarded $29,039,803 in federal dollars, attracting approximately $35 million in matching funds.

The federal share included $1,476,600 from unclaimed state funds (states that did not match their base funding). Appendix E lists the ranking criteria, reflecting the priorities listed on page 20. Table 2 summarizes the awards by region, and Appendix A contains a comprehensive list of the awardees and grant amounts.

The National Clean Diesel Emerging Technologies Program: Moving Innovations from Concept to Marketplace

EPA's NCDC has been demonstrating new technologies and creating new demands for environmentally superior products since its inception. To continue to encourage new technologies to meet the needs of fleets and air quality planners, Congress recognized the need for the DERA program to promote cutting-edge advanced technologies. EPA developed the National Clean Diesel Emerging Technologies Program to respond to Section 792 (b)(3)(B)(i), which directs that "...the Administrator shall provide not more than 10 percent of funds available for a fiscal year under this provision to eligible entities for the development and commercialization of emerging technologies." In FY 2008, under the National Diesel Emerging Technologies Program, 10 applicants requested \$5 million and offered more than \$1 million in matching funds. EPA awarded six grants in five states for a total of \$3.7 million in federal funds, with \$1.1 million in matching funds. These funds supported NO_X reduction technologies such as SCR systems for on-highway trucks and nonroad construction equipment, greater emission controls from a combination of diesel oxidation catalysts (DOCs)/DPFs and exhaust gas recirculation (EGR) systems, and new techniques for harbor craft, such as an emission upgrade kit for two types of marine engines. The criteria by which the applicants were evaluated are presented in Appendix E. The new federal dollars used for these emerging technologies are vital to spurring investment by NCDC's partners. The National Clean Diesel Emerging Technologies Program is an opportunity to advance new, cutting-edge technologies to reduce diesel emissions from the existing fleet. Building on EPA's verification program, EPA developed a process by which a manufacturer can request that its technology qualify in advance as an emerging technology. Through its collaborative network, EPA has been encouraging new technologies to reduce NO_X, save fuel, and perform better on a wider array of conditions and vehicles, including nonroad equipment and marine vessels.

Emerging Technology

An emerging technology is defined in Section 791 of the Energy Policy Act to be "... a technology that is not certified or verified by the Administrator or the California Air Resources Board but for which an approvable application and test plan has been submitted for verification to the Administrator or the California Air Resources Board."

To qualify as an emerging technology, manufacturers should be in the initial stages of the verification process with EPA or CARB and listed on EPA's Emerging Technology List.

The manufacturer must also provide an approvable application for verification (including an explanation of the engineering principles of the technology and why the technology should be considered an emerging technology) and an approvable test plan. An emerging technology must be close to being, if not already, commercially available.

Emerging Technology to Bring Large Reductions in Four Pollutants

EPA awarded a grant to the South Coast Metropolitan Air Quality Management District, in California, home to approximately 16 million people, to install selective catalytic reduction technology (SCRT) on 1999 to 2002 heavy-duty Class 8 on-road diesel trucks, which will reduce PM, NOX, CO, and HC emissions by at least 90, 65, 85, and 90 percent, respectively.

This project will improve the air quality of a high-population-density nonattainment area. At the same time, the project will produce useful data about the effectiveness and durability of this emerging technology.

Once approved, emerging technologies are listed on EPA's Web site. Technologies may remain on EPA's Emerging Technology List for up to two years. During that two-year period, EPA estimates that manufacturers will be able to complete the necessary steps to obtain full verification.

Once a technology is verified or certified, it will no longer be considered "emerging" but will then be fully eligible for projects under the National Clean Diesel Funding Assistance Program and then deployed more widely.

This program will accelerate the pace of introduction of new diesel emission reduction strategies and will allow the DERA program to evolve to better serve the public.

The SmartWay Clean Diesel Finance Program: Supporting Low-Cost Loans for High-Value Technologies

Under the SmartWay Clean Diesel Finance Program, EPA received four applications requesting $9.5 million and leveraging $44 million. Ultimately, in September 2008, EPA awarded three competitive grants totaling $3.4 million to establish a national financing program that provides funding to small- and medium-sized trucking companies to purchase clean diesel technologies, including leveraging more than $19 million in additional funding.

These innovative financing programs will help smaller trucking firms lower their emissions, fuel costs, and carbon footprint by purchasing cleaner used trucks equipped with diesel exhaust filters and/or fuel-saving technologies. The program will also support idling and emission reduction technologies for 2,460 trucks. Appendix E presents the criteria by which the applicants were evaluated. Appendix C provides additional details about the projects.

DERA provided EPA with new authority to establish national grant and low-cost revolving loan programs to finance clean diesel projects. Administered by EPA's SmartWay Transport Partnership, the FY 2008 RFP for this program focused on establishing low-cost loan programs for retrofitting used pre-2007 highway vehicles and new or used pieces of nonroad equipment with EPA- or CARB-verified emission control and idle reduction technologies. EPA encouraged financing proposals to include strategies such as loan guarantees, equity investments that leverage additional funds, tax-exempt or taxable bonds to create a low-cost loan program, and revolving loan funds.

Using up to 10 percent of the funds available for the national program, the SmartWay Clean Diesel Finance Program is designed to allow eligible entities to create financing mechanisms to assist fleets with the purchase of EPA-approved fuel-saving and emission reduction technologies. To date, EPA has offered grantees the opportunity to develop innovative financing projects instead of issuing direct federal loans.

Loan programs are important because they enhance financial sustainability and allow the funds to be used multiple times as the loans are repaid and additional loans can be offered to purchasers of cleaner vehicles or technologies. They are able to leverage substantial nonfederal funds to increase the federal dollar's effectiveness. For example, in the case of a finance program using an equity investment, the award funds could be used to leverage additional funds from a lending institution to increase the amount available for loans.

Although potentially far reaching, this program presents unique challenges that are not faced in the National Clean Diesel Funding Assistance Program. Unlike public school bus fleets, where the fleet owners have a strong motivation to reduce emissions, most private diesel truck and equipment owners do not have the same incentives and may not be as receptive to installing diesel emission reduction technologies.

EPA has attempted, in previous demonstration grants, to provide retrofit technologies to truck operators and has received only minimal interest.

Grant Offers Low-Interest Loans for SmartWay Upgrades

Using new authority to issue grants for loans and other innovative financing mechanisms, EPA awarded $1.13 million to the Community Development Transportation Lending Services, Inc. to establish a revolving loan fund for the purchase of used trucks that have SmartWay upgrades or will receive SmartWay upgrade kits.

The program uses the grant to provide a lower interest rate than borrowers would otherwise be able to get—in this case, 5.5 to 8.5 percent, based on each applicant's business history, available collateral, cash flow, credit score, and other factors.

The loan payback period is three to six years.

SmartWay℠
Transport Partnership
U.S. Environmental Protection Agency

To help manage these challenges, EPA's SmartWay program worked with a variety of stakeholders in the trucking industry including truckers, truck dealers, truck finance companies, and trucking associations. EPA concluded that the best incentive to install retrofit technologies on older trucks is to provide lower cost financing to purchase newer trucks already equipped with these technologies. In fact, the assessment shows that truckers will purchase cleaner diesel trucks equipped with advanced air pollution control devices if the monthly finance charge is $100 to $200 less than traditional trucks without these technologies. The federal grant funds can be used to lower the monthly finance cost by either extending the terms of the loans or reducing the interest rates.

The State Clean Diesel Grant and Loan Program: Reaching Its Goal of Programs in All 50 States

As a telling indication of the need for and interest in diesel emission reductions throughout the United States, all 50 states elected to participate in the FY 2008 State Clean Diesel Grant and Loan Program.

EPA apportioned a base amount of $196,880 to each state.[8] In addition, as another sign of the nationwide support for diesel emission reduction programs, 32 states were able to match the base funding, providing more than $6.3 million, and therefore received additional incentive funding, as allowed by DERA. For these states, EPA established a total grant of $492,200 (with $295,320 in federal funds and at least $196,880 in state funds).

With this money, each state established a clean diesel program within its air agency and worked in consultation with EPA to develop an effective work plan and application. The projects varied and utilized a number of emission reduction strategies.

A total of 21 states concentrated their efforts on school buses; nine focused on the construction sector; and other states focused on agriculture, transit buses, marine ports, rail, or trucking.

Several states that have never initiated a state clean diesel program before are starting new programs. Appendix D presents a list of projects.

EPA issued a *Federal Register* notice (73 FR 12728; March 10, 2008) that announced the funding opportunity and explained the main provisions of the State Clean Diesel Grant and Loan Program.

The state program is not a competition but, rather, an allocation process in which the states indicate their interest in participating and applying for the funds, and EPA awards a specific allocation based on the total number of states whose applications are approved.

Any state wishing to participate receives funding if the state submits an approvable work plan. Early on, the EPA Administrator sent a letter to the governor of each state out-lining the goals of the State Clean Diesel Grant and Loan Program.

Provisions for State Uses of State Clean Diesel Grant and Loan Program Funds

- States had significant flexibility in their proposed work plan.
- States were encouraged to use funds for EPA- and/or CARB-verified or certified technologies, although there were cases where emerging technologies or other cost-effective strategies were allowed, as they advanced the goals of the program.
- States could use funds to select various state fleets or projects for funding.
- States were encouraged not to use funds for emission reductions that are mandated under federal, state, or local law (similar to Section 792 (d)(2), "Use of Funds—Regulatory Programs").
- States could use up to 15 percent for administrative costs.

Inaugural Year Results in Impressive, Cost-Effective Reductions across All Sectors

Receiving significant levels of funding changes the complexion of possibilities for communities struggling with ways to reduce harmful diesel pollution. Instead of small demonstrations, entire fleets can take advantage of proven technologies. Federal investments can attract private capital through low-cost loans under the new DERA provisions. Financially strapped states can move forward to maximize public health programs and improve air quality for their citizens.

In the first year of DERA funding, EPA's programs achieved what they set out to do. More than 14,000 diesel-powered vehicles and pieces of nonroad equipment became cleaner as an immediate and direct result of this work.

DERA Funding Supported Reductions Across All Sectors

As shown in Figure 6, the FY 2008 grants supported a wide array of projects in numerous sectors, from agriculture and construction to marine and seaports to rail to transit. Collectively, the power of many small actions across the range of commercial, municipal, and private endeavors yielded significant emission reductions.

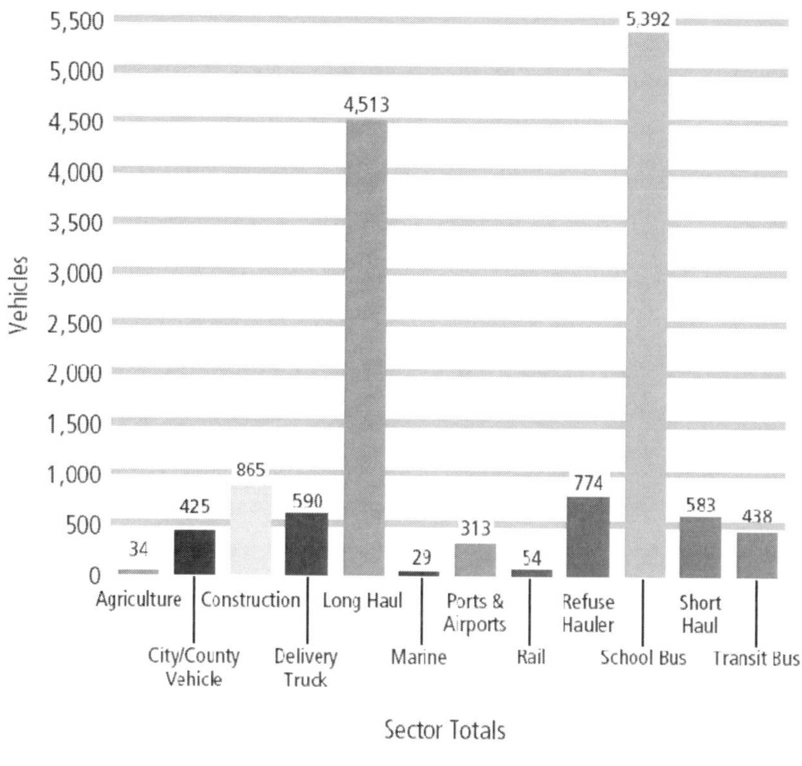

Figure 6. FY 2008 National Grants—Types and Numbers of Vehicles and Equipment.

DERA Funding Supported a Wide Range of Technologies and Fuels

As shown in Figure 7, the FY 2008 projects represent a wide variety of verified technologies, cleaner fuels, and certified engine configurations, such as repowers, replacements, idle reduction technologies (e.g., auxiliary power units [APUs], shore power, truck stop electrification), biodiesel, alternative fuels, lower-sulfur fuels in nonroad equipment, and retrofit devices such as diesel oxidation catalysts and diesel particulate filters.

> **State of Delaware Expands Its Program to Cover More Municipal Vehicles**
>
> EPA awarded the Delaware Department of Natural Resources and Environmental Control with a state clean diesel grant to retrofit 1998 Class 7 refuse haulers. Because the state put some of its own resources toward the program, EPA allocated additional incentive funding as well.
>
> This project is an excellent example of how the DERA funds allocated toward state programs can benefit the air quality of each state that opts into the program.

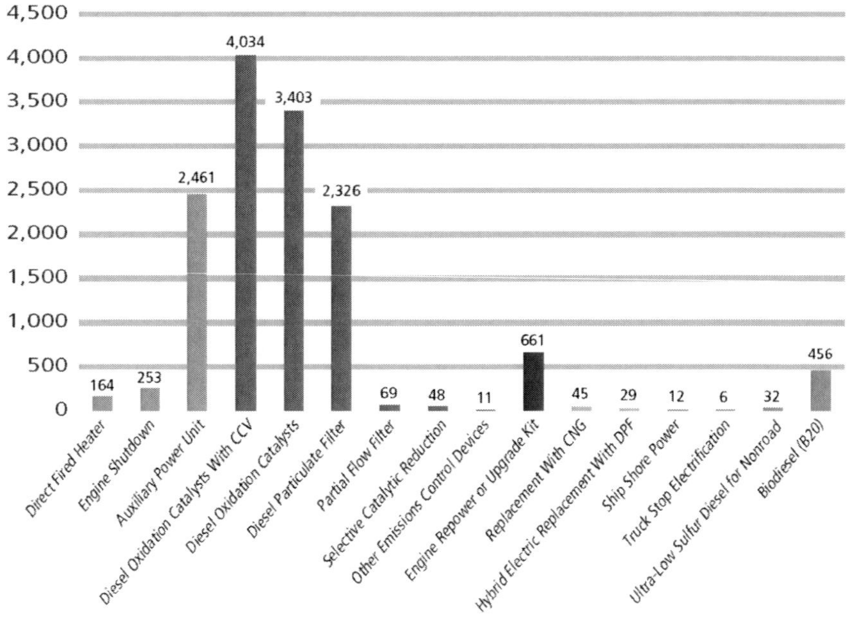

Figure 7. Technologies Employed in FY 2008 DERA Grants.

DERA Funding Helped Reduce the Amount of Diesel Fuel Used

Table 3 shows the impressive amount of diesel fuel conserved by the idle reduction programs DERA funded.

Taken together, the FY 2008 NCDC grants will save more than 3.2 million gallons of fuel per year, which corresponds to reducing 35,600 tons of CO_2 per year.

This savings equates to more than $8 million a year in fuel savings for fleet operators (assuming $2.50 per gallon fuel).

For comparison, freight trucks and locomotives consume 35 billion gallons of diesel fuel annually. Burning this fuel produces more than 350 million metric tons of carbon dioxide annually.

Table 3. Diesel Fuel Conserved Annually by FY 2008 DERA Projects With Idle Reduction Technologies

Program	CO_2 (annual tons)	Diesel Conserved (annual gallons)
National Clean Diesel Funding Assistance Program	8,200	736,000
National Clean Diesel Emerging Technologies Program	—	—
SmartWay Clean Diesel Finance Program	18,700	1,683,100
State Clean Diesel Grant and Loan Program	8,800	788,800
Total	**35,600**	**3,207,900**

Note: Values were rounded to the nearest 100 or to one significant figure. Sums may not add due to rounding. Calculation of CO_2 does not include CO_2 savings from engine repowers, replacements, or upgrades.

DERA Funding Results Are Relatively Cost-Effective

In addition to being successful in meeting their core goal, which is to reduce emissions, EPA's clean diesel programs are also cost-effective uses of federal funds, as shown in Table 4.

As NCDC has developed over the years, clean diesel projects have become more cost-effective as the technology has developed, as more vendors are competing, and as stakeholders have become more experienced in implementing projects.

In a series of studies, EPA analyzed clean diesel strategies generally and evaluated the cost-effectiveness of using add-on retrofits on buses, trucks, and nonroad equipment using a standard set of assumptions.[9]

Specifically, in these studies, EPA evaluated the cost-effectiveness of retrofitting school buses, freight trucks, and bulldozers with DOCs and catalyzed DPFs, two of the most common $PM_{2.5}$ emission reduction technologies for diesel engines.

Table 4. Projected Cost-Effectiveness

Program	Projected NO_X	Projected $PM_{2.5}$	Projected HC	Projected CO	Projected CO_2*	Program Federal Funding	Recipient Matching Amount
Federal costs per lifetime ton							
National Clean Diesel Funding Assistance Program	$2,000	$27,700	$15,300	$4,000	$300	$29,039,803	$34,749,839
SmartWay Clean Diesel Finance Program	$400	$9,000	$7,300	$2,100	$10	$3,390,000	$19,100,000**
State Clean Diesel Grant and Loan Program	$600	$16,700	$6,600	$2,300	$100	$12,994,080	$6,300,160
National Clean Diesel Emerging Technologies Program						$3,776,117	$1,130,927
Federal grant dollars per lifetime ton for three programs combined	**$1,000**	**$20,600**	**$10,500**	**$3,100**	**$100**		
Grant costs* per lifetime ton**							
Total dollars per lifetime ton* for three programs combined**	**$1,600**	**$33,300**	**$16,900**	**$5,000**	**$200**		

Note 1: Values were rounded to the nearest $100 or to one significant figure. Sums may not add due to rounding. Lifetime tons are not discounted.

* CO_2 was calculated for idle reduction technologies only and does not include reductions for repowers or replacements.

** Leveraged amounts; no matching was included in the grants, and this dollar amount is not included in the figure below.

*** Where cost includes federal funding, plus match, minus administrative costs assumed to be 15 percent, excluding leveraged amounts and the Emerging Technologies program.

Note 2: The Emerging Technologies program is designed to demonstrate new technologies. At this phase in the development of these technologies, costs would be expected to be higher than at a more fully mature commercial scale of deployment stage. Therefore, the costs comparisons with the other national program components shown in Table 4 have been omitted.

- For nonroad equipment retrofit device cost-effectiveness:
  - PM: DOCs and DPFs for nonroad equipment ranged from \$18,700 to \$87,600 per ton of $PM_{2.5}$ reduced.
  - NO_X: SCR systems and engine upgrade kits were calculated to range from \$1,900 to \$19,000 per ton of NO_X reduced.
- For on-highway applications, such as school buses and trucks cost-effectiveness:
  - PM: DOCs ranged from \$11,100 to \$67,700 per ton of $PM_{2.5}$ reduced and DPFs retrofits ranged from \$12,100 to \$69,900 per ton of $PM_{2.5}$ reduced.
  - NO_X: Truck stop electrification ranged from \$1,400 to \$2,000 per ton of NO_X reduced, and APUs ranged from \$2,700 to \$3,500 per ton of NO_X reduced.

These estimates of cost-effectiveness for retrofit devices depend on a number of factors such as equipment activity, survival rates, effectiveness of the system, and costs.

For early retirements of vehicles and engine replacements or repowers, the new engines must meet EPA's more stringent regulations and thus are significantly cleaner. Cost-effectiveness estimates were made to support EPA's Clean Air Act regulatory programs.

- For nonroad vehicles:
  - PM: For EPA's Tier 1/Tier 2 standards, the cost-effectiveness is estimated at \$2,000 per ton of $PM_{2.5}$ reduced.
  - NO_X:
    - For EPA's Nonroad Tier 4 standards, the cost-effectiveness is estimated at \$1,010 per ton (NO_X + nonmethane hydrocarbons [NMHC]).
    - For the locomotive and marine standards, the cost-effectiveness is \$30 to \$190 per ton (NO_X + NMHC).

These PM-related original equipment manufacturer costs are comparable to or lower than those of installing retrofit technologies on an existing vehicle because of economies of scale in mass production and related factors. Depending on the age of the engine being retired, replacements and repowers can also improve fuel efficiency. Replacements can be cost-effective strategies for the oldest vehicles and vessels in the fleet, but there is often a significant financial barrier, because EPA grants only offset a portion of the overall cost of the vehicle. Several of the FY 2008 grants take advantage of the very cost-effective NO_X and PM reductions possible with locomotive and marine repowers.

Similarly, EPA anticipates that as emerging technologies become commercialized and more widely available, their cost-effectiveness too will be comparable to other verified technologies; however, at this early stage in their development, comparisons can be somewhat misleading, so they are not included in the analysis.

For comparison, for the PM NAAQS Regulatory Impact Assessment, for $PM_{2.5}$ reductions:

- The annualized cost for electrostatic precipitators for utility boilers was \$3,000 to \$15,000 per ton of $PM_{2.5}$ and \$1,000 to \$20,000 per ton for industrial boilers.
- The cost of fabric filters for these same units ranged from \$2,000 to \$100,000 per ton of $PM_{2.5}$.
- Area source controls' annualized costs are less than \$100 per ton of $PM_{2.5}$.

When analyzing the effectiveness of the federal plus matching costs of the FY 2008 DERA projects, EPA found that they were within the expected range of cost-effectiveness. The average full program cost of the national, state, and SmartWay finance programs is about \$33,300 per ton of $PM_{2.5}$ (including the matching amounts and assuming an upper limit on administrative costs of 15 percent, excluding leveraged amounts).

EPA analyzed the way in which federal dollars were leveraging other funding to achieve emission reductions. The reductions achieved over the lifetime of the FY 2008 DERA grants are being financed in different manners. In some cases, for instance, the federal share of the grant might pay for the entire cost of the retrofit device. In other cases, the federal share of the grant might provide a small amount, coupled with state incentive funds or private investment. In the SmartWay Clean Diesel Finance grants and in the other program grants that used the new loan provisions, relatively small federal investments on a per-unit basis attracted large private investment, especially where additional economic benefit in terms of fuel savings or the value of a newer vehicle was involved.

For the federal competitive national program:

- **National Clean Diesel Funding Assistance Program:** The average federal cost was estimated at \$27,700 per ton of $PM_{2.5}$ reduced over the lifetime of the project and just under \$2,000 per ton of NO_X reduced over the project lifetime.
- **SmartWay Clean Diesel Finance Program:** The average federal cost was estimated at \$9,000 per ton of $PM_{2.5}$ reduced over the lifetime of the project and \$400 per ton of NO_X reduced over the project lifetime. Federal funds leveraged significant private sector funding and fuel savings in this new approach.

For the State Clean Diesel Grant and Loan Program:

- The average federal cost was estimated at \$16,700 per ton of $PM_{2.5}$ and \$600 per ton of NO_X over the lifetime of the project. State program costs are lower than national program costs because many states have established programs that are generally more limited to specific lower-cost vehicle categories or have leveraged amounts not captured here.

For the three programs together:

- The grants will reduce approximately 46,000 tons of NO_X and 2,240 tons of $PM_{2.5}$ over the programs' lifetimes in addition to substantial reductions in HC and other pollutants.
- The weighted average federal cost is about \$25,000 per ton of $PM_{2.5}$ and \$1,500 per ton of NOX reduced over the lifetime of the program.

DERA Funding Is Yielding Tangible Public Health Benefits

The DERA programs reduce emissions that affect people's everyday lives, especially for sensitive populations such as children, the elderly, those with heart and lung disease, outdoor workers, and people who exercise outdoors.

The diesel emission reductions resulting from the FY 2008 grants for NO_X and PM will total approximately 46,000 and 2,200 tons, respectively, by 2031, yielding a variety of health benefits. Fewer emissions leads to less air pollution, resulting in fewer respiratory symptoms, reduction in the use of medication for asthma, and fewer visits to emergency departments or health care providers. Parents miss fewer work days caring for their children or their own respiratory symptoms.

- The health benefits from the FY 2008 grant projects will range from a net present value of $580 million to $1.4 billion (2006 dollars), assuming a 3-percent discount rate throughout the lifetime of the program.
- These benefits include an estimated 95 to 240 avoided premature deaths.[10]

Ozone-related health benefits, cancer risk, and other welfare effects were not quantified in this analysis.

4. Resources That Support the National Clean Diesel Campaign: Partners Fulfill Commitments and Measure Progress

While reducing diesel emissions is the first priority, producing technical and programmatic support materials for stakeholders is a necessity as well. To ensure the successful implementation of NCDC, EPA developed the Diesel Emissions Quantifier (DEQ) calculator; guidance documents; public outreach materials including a dynamic Web site, an idle reduction campaign, and a Scholastic *Magic School Bus* book; and a Web-based tool kit for state and local programs.

Diesel Emissions Quantifier

To help state and local governments, fleet owners, school districts, municipalities, contractors, port authorities, and others in estimating emission reductions and cost-effectiveness for clean diesel projects, EPA created an interactive, Web-based calculator tool called the DEQ. Users enter specific information about a fleet, and emission and cost-effectiveness calculations are made based on existing EPA models and guidance. The DEQ performs real-time calculations and estimates project-level emissions for $PM_{2.5}$, NO_X, HC, CO, and CO_2. To support users of this tool, EPA also created a user guide and held interactive training webinars.

Ultimately, the tool became a critical component of NCDC, as most grant applicants rely on it to estimate their emission reductions. Grant applicants are required to evaluate the quantifiable and unquantifiable benefits of the emission reductions of their proposed projects in each grant application, using a methodology approved by EPA or the National Academy of Sciences.

Agency Guidance for Calculating Cost-Effectiveness

EPA published two peer-reviewed technical papers analyzing the cost-effectiveness of reducing emissions from both on-highway and nonroad diesel engines. (The full reports are available at: www.epa.gov/otaq/diesel/publications.htm.)

- The Cost-Effectiveness of Heavy-Duty Diesel Retrofits and Other Mobile Source Emission Reduction Projects and Programs (EPA-420-B-07-006, May 2007).
- Diesel Retrofit Technology: An Analysis of the Cost-Effectiveness of Reducing Particulate Matter and Nitrogen Oxides Emissions From Heavy-Duty Nonroad Diesel Engines Through Retrofits (EPA-420-R-07-005, May 2007).

In these papers, EPA evaluated the costs and emission benefits of retrofitting school buses, freight trucks, and bulldozers with DOCs and catalyzed DPFs, two of the most common PM emission reduction technologies for diesel engines. The findings from these studies indicate that retrofits can be a cost-effective way to reduce air pollution compared with other EPA programs.

Guidance on State Implementation Plans

EPA published guidance to help state and local air quality and transportation planners satisfy three EPAct 2005 Section 795 provisions: 1) quantify the emission reductions from retrofitting diesel vehicles, engines, and equipment; 2) appropriately include the emission reductions from diesel retrofits in state implementation plans (SIPs) to help demonstrate progress toward, attainment of, or maintenance of NAAQS; and 3) appropriately include emission reductions from diesel retrofits in transportation conformity or general conformity.

- Diesel Retrofits: Quantifying and Using Their Benefits in SIPs and Conformity—Guidance for State and Local Air and Transportation Agencies (EPA-420-B-06-005, June 2006).
- Truck Guidance: Guidance for Quantifying and Using Long-Duration Truck Idling Emission Reductions in State Implementation Plans and Transportation Conformity (EPA-420-B-04-001, January 2004).
- Locomotive Guidance: Guidance for Quantifying and Using Long-Duration Switch Yard Locomotive Idling Emission Reductions in State Implementation Plans (EPA-420-B-04-002, January 2004).

Public Outreach Materials

EPA has created a robust national Web site (www.epa.gov/cleandiesel) and myriad public out-reach materials to assist partners and others with diesel emission reduction projects. In addition, each Regional Collaborative maintains a Web site with regional priorities and resources.

As a sign of the national Web site's utility, more than 80,000 users visited the site in FY 2008. The site is continually updated with vital new information and materials, tailored to meet stakeholder needs. Its design includes a searchable list of demonstration projects, case studies, technology information, listservs, a trucker portal (www.epa.gov/otaq/diesel/truckers. htm), sector-specific content (www.epa.gov/cleandiesel/ports), and other helpful links, including links to Regional Collaborative Web sites.

The breadth of EPA's reach in communicating information about clean diesel and NCDC to the public in FY 2008 can be measured by the following statistics:

- 100 diesel conferences supported by EPA via financial sponsorships, promotion, and participation, including the delivery of 12 major speeches by EPA's senior executives.

- 19 press releases issued highlighting clean diesel accomplishments and their benefit to the public.
- 40,125 program brochures distributed to stakeholders.
- 650 Clean School Bus USA idle reduction training videos in use.
- 300 Clean Construction USA videos viewed.
- 128,000 copies of Scholastic's *The Magic School Bus Gets Cleaned Up* books distributed.

NCDC Joins Forces with International Partners

Using funding from non-DERA sources, partnering with local governments and nongovernmental organizations, and supported by U.S. technology vendors, EPA led demonstration projects in Mexico City, Mexico; Bangkok, Thailand; Santiago, Chile; Beijing, People's Republic of China; and Pune, India.

Tool Kit for State and Local Government Partners

To support state, regional, and local governments in improving air quality and public health through diesel engine emission reduction efforts, EPA developed a Web-based tool kit that compiles examples of program components they could model (www.epa.gov/otaq/diesel/slt/basicinfo.htm). The content focuses on designing programs, communicating success, identifying funding, and evaluating program results. This tool kit is just one example of how NCDC operates as a shared responsibility between EPA and state and local air quality agencies.

EPA retrofitted about 20 vehicles in each city to demonstrate the emission reductions obtainable with various devices. The projects helped educate government and nongovernment staff about clean diesel technologies and aim to build the capacity so they can conduct retrofits on their own. Following the Beijing demonstration project, the city of Beijing independently retrofitted more than 5,000 diesel vehicles in preparation for the 2008 Summer Olympics. EPA also worked with the U.S. Department of Transportation's Maritime Administration on a project with the Port of Shanghai that resulted in a broader Pacific Ports Initiative, promoting cleaner diesel technologies among key trading partners and their ports. Also on the international front, EPA launched a cooperative project among the Port of Rotterdam, the Port of New York and New Jersey, a terminal operator with facilities in both ports, and technology providers to develop hydraulic hybrid cargo handling equipment. This technology promises to dramatically improve fuel economy while reducing PM and NO_X. Lastly, EPA hosted a workshop for Canadian federal and provincial staff in June 2008. The workshop transferred knowledge and built relationships with appropriate Canadian government personnel so that they might learn from the NCDC efforts and be better equipped to launch their own program. These international actions are consistent with DERA Section 795(d), "Outreach and Incentives; International Markets."

5. Remaining Challenges : Adding Verified Technologies and Incentives Could Enhance Implementation

EPA's advance work in developing guidance and materials, offering educational tools to partners developing clean diesel programs, and communicating consistently and clearly with partners and stakeholders was essential to the success of the first year. At the same time, the Agency's ability to troubleshoot problems that arose and working within the collaborative and sector networks to improve and adjust the program along the way ensured that it met its mission. The challenges encountered in implementing the first year of this program were relatively few. At the end of the year, however, two challenges remain:

1. **The number of verified technologies for nonroad and marine engines is too few.**
   While the list of technologies available on EPA and CARB's lists for on-highway applications is long, the technologies available for nonroad and marine applications are limited. The agencies have

identified a lack of laboratory facilities and field testing instruments capable of evaluating marine technologies as the main barrier. Both agencies have been focusing their efforts on increasing the availability of nonroad retrofit technologies, closely working with technology manufacturers to encourage additional technology verification for nonroad and marine equipment. There are also some important gaps in diesel retrofit technologies for older heavy-duty diesel trucks. While EPA and CARB have verified numerous heavy-duty diesel truck retrofit technologies, the low-cost devices only reduce PM emissions by a limited amount, and the more effective devices might not work as well with the older trucks' duty cycles and business models. In some cases, the value of the retrofit device can exceed the remaining value of an old truck. Accordingly, EPA will continue to work with diesel retrofit manufacturers to identify cost-effective solutions for older diesel trucks. Responding to this need, EPA held a Retrofit Technology Verification Workshop in Washington, D.C., on December 13, 2007. The workshop was an overwhelming success, with nearly 100 people attending, including technology vendors, original equipment manufacturers, fuel and fuel additive makers, diesel engine technology organizations, and state and local governments such as CARB. This workshop spurred additional verification activity among manufacturers. In 2008, EPA approved seven additional technologies for nonroad applications and 13 additional technologies for on-highway applications. In addition, EPA approved six emerging technologies (three nonroad and three on-highway applications), which will continue in the verification process and could well be verified within the next two years.

Despite these gains, EPA had to remove a number of technologies from its Technology Verification List in December 2008, due to the implementation of a new, more stringent limit on the amount of unwanted nitrogen dioxide (NO_2) that retrofit devices are allowed to produce. EPA anticipates that some of the removed technologies will soon meet the new requirement and the Agency will be able to reinstate them.

2. **New incentives are needed to retire the oldest and dirtiest engines.** As EPA's Clean Diesel regulations under the Clean Air Act phase in, new engines that must meet significantly more stringent standards are increasingly available in the marketplace. The challenge is to create

additional incentives for private fleet owners to replace their older diesel engines with new engines that meet the new standards.

Under the current law, national program grants cannot go toward private fleet owners or to projects for which there is a federal, state, or local mandate. First, it is a challenge to encourage private fleets to apply, as they must partner with eligible entities, creating a barrier to participation especially for small businesses that may own or operate the oldest and dirtiest fleets. Second, routine attrition of old vehicles is considered to be compliance activity with Clean Air Act requirements and currently is not eligible for funding. Accordingly, early replacement projects can receive federal grants, but not projects that might target the oldest, dirtiest engines that are already past their useful life and should have been replaced through normal attrition but are nevertheless still in use. These dual challenges are especially great for small business owner/operators for equipment such as drayage trucks that serve seaports, tug and tow vessels, and construction equipment. EPA will continue to work to identify incentives utilizing the innovative finance and loan provisions of the statute and our SmartWay Transport Partnership, as well as to fashion additional incentives together with our state and local government partners.

6. Lessons Learned and Needs for the Future: Continue, Improve, and Expand

While the DERA program has achieved significant emission reductions from diesel engines in its first year, a tremendous amount of work remains, and substantial investments are needed.

Lessons Learned

Lessons learned from the first year include the following:

- **EPA and others must continue efforts to utilize existing national and regional networks to further improve the program results.** In order to expedite the program, EPA developed materials to communicate the program options, technologies, goals, and schedules.

Providing clear information and catalyzing networks allowed all parties to engage in the process and move projects forward. As the programs grow, EPA must promote efforts that allow agencies to share experiences to encourage action. To this end, EPA will work to strengthen existing regional diesel collaboratives.

- **EPA must continue to cultivate state and local leadership.** All 50 states elected to participate in the first year of the program. For some, this was just the beginning of their work in this program area. To ensure future success, EPA must continue to work with state agencies and the District of Columbia to build comprehensive clean diesel programs.

  The results of the program can be seen at the state and local levels where leadership from state and local agencies, nonprofit organizations, and private organizations engages interested fleets in the program and works with them to achieve results. EPA must encourage more organizations to take this role as both an environmental initiative and an economic opportunity.
- **EPA and other organizations must continue to involve the private sector.** The private sector plays a key role in implementing the program. Organizations that supply eligible technologies must continue to work to improve their technologies and identify opportunities for fleets that are willing to take action. In addition, EPA and other agencies should encourage private fleets in priority areas to take proper steps to address the air pollution from their fleets. The SmartWay Transport Partnership, for example, has created a program for private fleets in the freight sector.

  In exchange for making a commitment to reduce emissions and fuel usage, EPA provides technical advice, public recognition, and the opportunity to become preferred carriers for SmartWay shippers. Through corporate leadership to reduce emissions and reduce fuel usage, the DERA program has a great opportunity for expansion. In order to accomplish this, corporate champions must step forward and make measurable commitments.
- **Organizations must continue to develop and advance verified nonroad and marine technologies.** Opportunities continue to arise in new program areas as the program grows. Vendors must recognize these opportunities and work to create improved options to address the multitude of emission sources covered under the legislation, especially in the nonroad and marine sectors.

EPA will expand the emerging technology and verification programs to promote innovative technologies that reduce priority pollutants and conserve fuel.

While the verified technologies list available on EPA's and CARB's Web sites for on-highway applications is long, there are limited products verified for nonroad, marine, or older diesel trucks.

There are limited products available to reduce NO_X for any category of engine. Adequate laboratory facilities and field testing instruments capable of evaluating PM emission reductions from marine technologies will be needed.

EPA has been working closely with manufacturers to verify technologies and fill these gaps.

The nonroad market is complicated by the number and diversity of nonroad equipment types, by the range of horsepowers and engine types involved, and by the varying usage and duty cycles of the equipment. There are many opportunities, however, to provide cost-effective diesel emission reduction technologies to meet local air quality needs.

Next Steps

Moving forward in the program, EPA commits to the following:

- Continue to work aggressively to reduce pollution from diesel engines across the country by partnering with key stakeholders to promote clean diesel strategies.
- Target current PM and ozone nonattainment areas where clean diesel strategies will have the greatest public health impact.
- Provide assistance to state and local governments in developing their own clean diesel programs.
- Continue to provide high-quality data to states that depend on the performance of diesel emission reduction strategies in their air quality plans, through in-use testing.
- Continue to work cooperatively with CARB and other states to provide a robust list of clean diesel technology options for partners.
- Continue to confirm the emission performance of verified technologies in the field.
- Continue to develop innovative financing approaches for stretching federal dollars to achieve maximum diesel emission reductions,

especially where fuel savings can create a business case for low-cost loans or other financial incentives.

- Develop timely educational materials to continue to build awareness of clean diesel opportunities.

In summary, the first year of implementation for the Diesel Emissions Reduction Program has been extremely successful. By harnessing broad support from environmental organizations, state and local governments, industry, technology vendors, and other groups, EPA has been able to magnify the available dollars and strategically fund important emission reduction and fuel saving projects.

In total, EPA awarded 119 grants in the first year of the DERA program. More than 14,000 vehicles and pieces of nonroad equipment in a wide array of sectors will be cleaner as a direct result of this program. New clean diesel programs have been established in every state and technologies have been advanced for the future.

The implementation of the EPAct 2005's Diesel Emissions Reduction Program is a critical step toward reaching our national clean air goals and protecting public health and the environment for all Americans for generations to come.

APPENDIX A. NATIONAL CLEAN DIESEL FUNDING ASSISTANCE PROGRAM, FY 2008 GRANTS

State	Grant Recipient	EPA Award	Match	Sector(s)	Technology Types*
AL	Alabama State Port Authority	$750,000	$1,580,000	Ports, Rail	Replacement
AZ	City of Phoenix	$553,629	$732,000	Utility	Engine Replacement, DPF
CA	CALSTART, Inc.	$895,827	$1,015,325	Construction, Utility, Transit Bus	CNG, DPF, B20 Fuel
CA	Kern County Superintendent of Schools	$540,000	$540,000	School Bus	CNG Replacement
CA	Sacramento Metropolitan Air Quality Management District	$553,630	$2,581,600	Utility, School Bus	DPF, Engine Replacement
CA	South Coast Air Quality Management District	$1,000,000	$7,095,900	Long-Haul Trucks	DPF
CO	City and County of Denver	$200,000	$87,059	Utility, Construction	Heater
CO	Colorado Department of Public Health	$400,000	—	School Bus	DOC, Closed Crankcase Ventilation (CCV) System and Engine Preheater
CO	Regional Air Quality Council	$455,645	—	School Bus	DOC, CCV System
CT	Connecticut Department of Environmental Protection	$50,000	—	Utility, Construction	Diesel Oxidation Catalyst, DPF, Ultra-Low Sulfur Diesel
DC	Metropolitan Washington Council of Governments	$598,516	$193,983	Construction	DPF, Repower, Engine Upgrade
FL	City of St. Petersburg	$666,510	$5,745	Utility, Construction	DPF, B20 Fuel
GA	Georgia Ports Authority	$250,000	$33,075	Ports	DOC, CCV System
ID	Idaho Department of Environmental Quality	$500,000	$500,000	School Bus	DPF, Partial Flow Filter, DOC
IL	Chicago Public Schools	$373,909	$92,000	School Bus	Retrofit (DOC, DPF)

State	Grant Recipient	EPA Award	Match	Sector(s)	Technology Types*
IL	Illinois EPA	$678,604	$654,699	School Bus, Transit Bus, Utility	Retrofit (DOC, DPF, CCV System), Auxiliary Power Unit, Direct-Fired Heater, Hybrid Replacement
IN	Indiana Department of Environmental Management	$334,500	$61,257	Transit Bus	Retrofit (DOC, DPF), APU
IN	Northwest Indiana Forum Foundation Inc.	$164,032	$173,432	Construction	Repower
KS	Kansas Department of Health and Environment	$1,525,524	$1,871,500	Trucks and Transit Buses	DOC w/ CCV System
KY	Kentucky Clean Fuels Coalition	$473,939	$2,160,000	Ports	Vehicle Replacement
MA	Massachusetts Port Authority (Massport)	$400,000	$147,815	Ports	Shore Power
MA	NESCAUM	$400,000	$100,000	Construction	DPF
MA	NESCAUM	$535,250	—	Rail	Auxiliary Power Unit
MD	Maryland Environmental Services	$361,951	$25,909	Ports	DPF
MD	Montgomery County Public Schools	$699,501	$7,350	School Bus	DPF
MI	Lenawee Intermediate School District	$154,381	—	School Bus	DPF
MI	Michigan Clean Energy Coalition	$250,000	$250,580	Construction	Repower
MI	NextEnergy Center	$250,000	$650,500	Long Haul Trucks	APU
MI	Laker School District	$251,100	$253,220	School Bus	Bus Replacement; Idle Reduction Technology, B20 Fuel, Heater
MN	Minnesota Environmental Initiative	$400,000	—	School Bus, Utility	Retrofit (DOC, CCV System)
MO	Grace Hill Settlement House	$454,849	—	School Bus	Crankcase Filtration System (CFS)
MO	Missouri Department of Natural Resources	$726,227	$621,550	Construction	DOC, CCV System, APU, Engine Shutdown
NC	North Carolina Department of Environment and Natural Resources	$750,000	—	Construction	Repower, Replacement

(Continued)

State	Grant Recipient	EPA Award	Match	Sector(s)	Technology Types*
NH	Manchester Transit Authority	$229,703	$5,652	School Bus, Construction, Utility	DOC, CCV System, S-Bar Heater, B20 Fuel
NJ	New Jersey Motor Truck Association	$503,285	—	Long-Haul Trucks	APU, Bunk Heater, DOC, DPF
NY	Board of Cooperative Educational Services of Ulster County	$130,690	—	School Bus	Vehicle Replacement
NY	Capital District Transportation Authority	$125,000	—	Transit Bus	DPF
NY	Environmental Defense Fund	$400,000	$2,025,000	Utility	Vehicle Replacement (Hybrid-Electric)
NY	Erie County Department of Environmental Planning	$523,395	$72,278	School Bus	DOC, CCV System, Diesel-Fired Engine Warm-Up Heater
NY	Middle Country Central School District	$359,305	$361,305	School Bus	Vehicle Replacement (CNG)
NY	Scarsdale Union Free School District	$346,240	$29,020	School Bus	DOC, CCV System
NY-NJ	Port Authority of New York and New Jersey (PANYNJ)	$280,500	$80,500	Utility	DPF, Flow Through Filter, DOC
NY-NJ	PANYNJ	$750,000	$1,250,000	Ports	Flow Through Filter, DPF
OH	Ohio Environmental Council	$394,589	$29,571	Ports, Utility, Refuse Haulers	Retrofit (DOC, DPF, CCV System); APU
OH	Ohio Indiana Clean Diesel Collaboration	$412,554	$81,344	Utility	DPF
OH	Stark County Educational Service Center	$465,364	$89,905	School Bus	Retrofit (DPF, CCV System, DOC), Replacement
OR	City of Portland	$498,726	$58,658	Ports, Construction	DOC, Exhaust Retrofit, Ultra-Low Sulfur Diesel
PA	City of Philadelphia	$750,000	$2,250,000	Utility	CNG Vehicle Replacement
PA	Pennsylvania Department of Transportation	$219,434	$24,381	Rail	APU

State	Grant Recipient	EPA Award	Match	Sector(s)	Technology Types*
SC	South Carolina Port of Charleston	$735,002	$963,502	Utility, Construction	Engine Repower, Ultra-Low Sulfur Diesel
SD	Sioux Falls School District	$300,000	$300,000	School Bus	Replacement
TN	Knox County Government	$163,871	$23,056	Utility, Construction, Agricultural	Emission Control, Ultra-Low Sulfur Diesel
TX	North Central Texas Council of Governments	$750,000	$750,000	Trucks	APU
TX	North Central Texas Council of Governments	$750,000	$1,500,000	Construction	Replacement
TX	Texas Commission on Environmental Quality	$500,000	—	School Bus	DOC, DPF, Partial Flow Filter, CCV System
UT	Utah Department of Air Quality	$400,000	$4,000	School Bus	DOC and CCV System
VA	Port of Norfolk	$647,457	$1,732,600	Ports and Rail	Engine Repower
VT	Chittenden Solid Waste District	$205,523	$616,569	Utility	Vehicle Replacement
WA	Puget Sound Clean Air Agency	$850,000	$318,000	Ports	Vehicle Replacement, DPF, Partial-Flow Filter, Crankcase Ventilation Filter, DOC
WI	Wisconsin Department of Transportation	$750,000	$750,000	Construction	Repower
Total		**$29,038,162**	**$34,749,839**		

* Refer to page 56 for a list of acronyms and abbreviations.

Appendix B. National Clean Diesel Emerging Technologies Program, FY 2008 Grants

State	Grant Recipient	EPA Award	Match	Sector(s)	Vehicle Types	Technology Types
CA	South Coast Metropolitan Air Quality Management District	$900,000	$858,614	Delivery Truck	Class 8a (33,001–60,000 lbs)	Selective Catalytic Reduction (SCR)
MD	Montgomery County Maryland	$400,000	$88,168	Delivery Truck	Class 6 (19,501–26,000 lbs)	SCR
TX	Center for Transportation and the Environment	$300,000	—	Short Haul	Class 8a (33,001–60,000 lbs)	Other Emissions Control Devices
TX	Texas Transportation Institute	$500,000	$64,200	Construction	Graders	SCR
TX	University of Houston	$500,000	$119,945	Construction	Tractors, Loaders, Backhoes	SCR
WA	Puget Sound Clean Air Agency	$700,000	—	Marine	Marine	Engine Upgrade
Total		**$3,300,000**	**$1,130,927**			

For the FY 2008 National Clean Diesel Emerging Technology Program competition, EPA approved six technologies, as shown below. These included two technologies for marine vessels and SCR technologies for on-highway vehicles and nonroad equipment that expand technological options to reduce NO_X by approximately 65 percent. This list of EPA's emerging technologies is updated as information is submitted from manufacturers, and EPA anticipates that the types and numbers of technologies will continue to grow.

Manufacturer	Technology	Application	Reductions (%)			
			PM	CO	NO_X	HC
Caterpillar, Inc.	Engine Upgrade Kit	Marine	25	N/A	N/A	N/A
ESW Canada	DOC, CCV System	Marine	25	70	N/A	25
Johnson Matthey	Urea-Based SCR System, DPF	On-Highway	90	85	65	95
Nett Technologies, Inc.	Urea-Based SCR System	Nonroad	20	60	65	60
Tinnerman/Shadowood	Reformer, LNT, SCR, DPF	On-Highway	90	90	65	90
Truck Emission Control Technologies Inc.	DOC, DPF, EGR	On-Highway	50	70	40	60

APPENDIX C. SMARTWAY CLEAN DIESEL FINANCE PROGRAM, FY 2008 GRANTS

Grant Recipient	Project Overview
Cascade Sierra Solutions Everybody Wins USA Lease Program Grant Amount: $1.13 million	**Loan Interest Rate:** 8% to 11%, includes 7% for financing cost, 1% collection fee, and 1% to 3% reservation loss. **Loan Payback Period:** Three years. **Eligible Activities:** Installation and leasing of EPA SmartWay or CARB-verified emission and idle reduction technologies. The truck owners will purchase the equipment for $10 at the end of the loan period. **Program Goal:** Establish a leasing program for the installation of emission reduction technology and idle reduction technology on 1,700 trucks.
Community Development Transportation Lending Services, Inc. National Low Interest Revolving Loan Fund	**Interest Rates:** 5.5% to 8.5%, varies based on, among other things, each applicant's business history, available collateral, cash flow, and credit score.

(Continued)

Grant Recipient	Project Overview
Grant Amount: $1.13 million	**Loan Payback Period:** Three to six years, varies based on, among other things, business history, available collateral, and cash flow. **Eligible Activities:** Purchase used trucks to upgrade with EPA SmartWay or CARB-verified idle reduction and/or emission reduction technology. Purchase used trucks that are upgraded with EPA SmartWay or CARB-verified idle reduction and/or emission reduction technology. **Program Goal:** Establish a revolving loan fund for the purchase of used trucks that have SmartWay upgrades or will receive SmartWay upgrades. **Service Fee:** 2% to 3%.
Owner-Operator Independent Drivers Association Foundation Innovative Financing Grant Amount: $1.13 million	**Loan Interest Rate:** Varies based on market Solid Waste Alternative Program rate. **Eligible Activities:** Purchase and installation of emission reduction and idle reduction technologies verified by EPA and/or CARB. **Program Goal:** Offer rebates to selected applicants/truck owners to install emission reduction and/or idle reduction equipment on their trucks. The first rebate to the truck owner will be for the cost of the equipment installation. Then, every six months, depending on idle/fuel use reductions demonstrated, the truck owner will receive additional rebates until 40% of the total cost of the equipment is paid.

APPENDIX D. STATE CLEAN DIESEL PROGRAM, FY 2008 GRANTS

State	Grant Recipient	Total Federal Funding Allotment	Sector(s)	Technology Types
AK	Alaska Department of Environmental Conservation	$295,320	Construction	Research, Retrofit
AL	Alabama Department of Environmental Management	$196,880	Rail	Repower
AR	Arkansas Department of Environmental Quality	$295,320	School Bus, Long Haul Trucks, Construction	Retrofitted, Idle Reduction
AZ	Arizona Department of Environmental Quality	$196,880	Short Haul	Truck Stop Electrification

State	Grant Recipient	Total Federal Funding Allotment	Sector(s)	Technology Types
CA	California Air Resources Board	$295,320	Long Haul Trucks	Retrofit
CO	Colorado Department of Public Health and Environment	$196,880	School Bus	Retrofit
CT	Connecticut Department of Environmental Protection	$295,320	School Bus, Long Haul Trucks	Idle Reduction
DE	Delaware Department of Natural Resources and Environmental Control	$295,320	Refuse Hauler	Retrofit
FL	Florida Department of Environmental Protection	$295,320	Long Haul Trucks, School Bus	Retrofit, Idle Reduction
GA	Georgia Department of Natural Resources	$295,320	School Bus, Construction	Retrofit
HI	Environmental Health Administration	$196,880	School Bus, Utility	Retrofit
IA	Iowa Department of Natural Resources	$196,880	School Bus	Replacement
ID	Idaho Department of Environmental Quality	$196,880	School Bus	Retrofit
IL	Illinois Environmental Protection Agency	$295,320	Long Haul Trucks	Idle Reduction
IN	Indiana Department of Environmental Management	$295,320	Long Haul Trucks	Retrofit
KS	Kansas Department of Health and Environment	$295,320	Long Haul Trucks	Replacement, Idle Reduction
KY	Kentucky Division for Air Quality	$196,880	School Bus	Retrofit
LA	Louisiana Department of Environmental Quality	$295,320	Marine	Repower
MA	Massachusetts Department of Environmental Protection	$295,320	Refuse Hauler, Rail, Utility	Retrofit, Replacement
MD	Maryland Department of the Environment	$295,320	Transit Bus	Retrofit
ME	Maine Department of Environmental Protection	$295,320	School Bus, Long Haul Trucks	Idle Reduction
MI	Michigan Department of Environmental Quality	$295,320	School Bus	Retrofit
MN	Minnesota Pollution Control Agency	$295,320	School Bus, Long Haul Trucks	Retrofit, Idle Reduction
MO	Missouri Department of Natural Resources	$295,320	Long Haul Trucks, Utility	Replacement, Idle Reduction
MS	Mississippi Department of Environmental Quality	$295,320	Utility Vehicle, School Bus	Retrofit
MT	Montana Department of Environmental Quality	$295,320	School Bus, Transit Bus	Retrofit, Repower

(Continued)

State	Grant Recipient	Total Federal Funding Allotment	Sector(s)	Technology Types
NC	North Carolina Department of Environment and Natural Resources	$295,320	Transit Bus, School Bus, Rail, Construction, Refuse Haulers	Retrofit, Idle Reduction, Repower
ND	North Dakota Department of Health	$196,880	School Bus	Replacement
NE	Nebraska Department of Environmental Quality	$196,880	School Bus, Refuse Hauler, City/County Vehicle, Transit Bus, Construction	Retrofit
NH	New Hampshire Department of Environmental Science	$196,880	Transit Bus, Construction, Utility	Retrofit
NJ	New Jersey Department of Environmental Protection	$295,320	Construction, School Bus	Retrofit
NM	New Mexico Environment Department	$196,880	School Bus	Retrofit
NV	Nevada Division of Environmental Protection	$295,320	School Bus	Retrofit
NY	New York State Department of Environmental Conservation	$295,320	Marine	Replacement
OH	Ohio Environmental Protection Agency	$295,320	School Bus	Retrofit
OK	Oklahoma Department of Environmental Quality	$295,320	School Bus	Retrofit
OR	Oregon Department of Environmental Quality	$295,320	School Bus	Retrofit
PA	Pennsylvania Department of Environmental Protection	$295,320	Ports, Airports	Repower
RI	Rhode Island Department of Environmental Management	$196,880	Ports, Airports	Retrofit
SC	South Carolina Department of Health and Environmental Control	$295,320	School Bus, Construction	Retrofit, Repower
SD	South Dakota Department of Environment and Natural Resources	$196,880	School Bus	Retrofit
TN	Tennessee Department of Environment and Conservation	$196,880	Long Haul Trucks	Idle Reduction
TX	Texas Commission on Environmental Quality	$295,320	School Bus	Retrofit

State	Grant Recipient	Total Federal Funding Allotment	Sector(s)	Technology Types
UT	Utah Department of Environmental Quality	$295,320	School Bus	Retrofit
VA	Virginia Department of Environmental Quality	$196,880	Utility Vehicle	Retrofit
VT	Vermont Department of Environmental Conservation	$196,880	School Bus	Retrofit, Idle Reduction
WA	Washington State Department of Ecology	$295,320	School Bus	Retrofit
WI	Wisconsin Department of Natural Resources	$295,320	Refuse Hauler, Construction	Retrofit, Repower
WV	West Virginia Department of Environmental Protection	$196,880	Transit Bus	Replacement
WY	Wyoming Department of Environmental Quality	$196,880	School Bus	Repower
Total		**$12,994,080**	**Match**	**$6,300,160**

APPENDIX E. NATIONAL PROGRAM EVALUATION CRITERIA

For the National Clean Diesel Funding Assistance Program, all grant competitions used criteria and points similar to those summarized in the table below, consistent with the priorities described in Section 792 of the Energy Policy Act, Subtitle G (see page 20), and with Agency policy.

National Clean Diesel Funding Assistance Program

For the National Clean Diesel Funding Assistance Program, the seven EPA Regional Collaboratives described in Appendix F held grant competitions, following Agency policy. Regions evaluated each Regional RFP following the same general criteria, supplementing with regional priorities worth 10 points in the final criterion.

EPA devoted the majority of the funding available to the National Clean Diesel Funding Assistance Program and allocated it to its 10 regional offices to award as competitive grants in support of clean diesel projects. EPA distributed funds across the regions based on a funding formula that included a base amount and considered population, diesel emission inventories, and other factors consistent with the objectives listed in the statute Section 792.

EPA's regions addressed certain regional priorities in their competitions, while following a model RFP to ensure that all of EPA's actions adhered to the statutory requirements and EPA policy.

National Clean Diesel Emerging Technologies Program

For the National Clean Diesel Emerging Technologies RFP, EPA employed similar criteria to those in the table below; however, EPA required that the projects propose to use technologies on EPA's Emerging Technology List by the closing date of the RFP. Criterion 9 was not applicable.

SmartWay Clean Diesel Finance Program

For the SmartWay Clean Diesel Finance RFP, EPA used criteria similar to those in the table below; however, Criterion 9 was not applicable.

Criteria	Points
1. **Project Summary/Approach:** EPA will evaluate the quality and extent to which the narrative proposal includes a well-conceived strategy for addressing the program priorities from the statute and a well-conceived, logical strategy for achieving the anticipated environmental results associated with the proposed project in a timely manner.	25
2. **Programmatic Priorities:** The Agency will evaluate the quality and extent to which the pro-posed project addresses the programmatic priorities.	10
3. **Past Performance—Programmatic Capability and Reporting on Environmental Results:** The Agency will evaluate the applicant's technical ability to successfully complete and manage the proposed project, taking into account the applicant's past performance in successfully completing and managing federally funded assistance agreements.	10
4. **Environmental Results—Outcomes and Outputs:** The Agency will evaluate the effectiveness of the applicant's plan for tracking and measuring its progress toward achieving expected project outputs and outcomes.	15
5. **Budget/Resources:** The Agency will evaluate whether the proposed project budget is appropriate to accomplish the proposed goals and measurable environmental outcomes.	10

Criteria	Points
6. **Clear Description of the Target Fleet:** Applicants will be evaluated on the degree to which detailed information on the fleet (vessel[s], vehicle[s], and/or equipment) is provided on the Applicant Fleet Description Spreadsheet and the eligibility of technology for the fleet.	10
7. **Leveraging Resources and Partnering:** Applicants will be evaluated based on the extent to which they demonstrate how they will coordinate the use of EPA funding with other federal and/ or nonfederal sources of funds to leverage additional resources to carry out the proposed projects and/or how that EPA funding will complement other activities.	15
8. **Staff Expertise/Qualifications:** Applicants will be evaluated on staff expertise/qualifications, staff knowledge, and resources or the ability to obtain them, to successfully achieve the goals of the proposed project.	5
9. **Regional Significance:** For the National Clean Diesel Funding Assistance Program, a factor for regional significance was added.	10

Appendix F. EPA Regional Clean Diesel Collaboratives

Beginning in 2003, EPA's regions began organizing Regional Clean Diesel Collaboratives with their states, local nonprofit organizations, private industry, and municipalities. The collaboratives listed below play a vital role in EPA's incentive-based approach to achieving superior environmental results. By tying into a network of regional stakeholders, this collaborative structure is ideal for achieving significant emission reductions across a large geographic area. Regional initiatives and state programs provide immediate and significant environmental results by working collaboratively with businesses, government and community organizations, industry, and others.

Northeast Diesel Collaborative

www.northeastdiesel.org

The Northeast Diesel Collaborative is a local initiative that builds on a foundation of voluntary action. Developed with EPA Regions 1 and 2, the initiative encourages participants to engage in projects that will reduce transportation-related air pollution to help address the high asthma rates in the Northeast. The program partners with the Northeast States for Coordinated Air Use Management (NESCAUM).

Mid-Atlantic Clean Diesel Collaborative

www.dieselmidatlantic.org

The Mid-Atlantic Clean Diesel Collaborative is a partnership among leaders from federal, state, and local governments; the private sector; and environmental groups in Delaware, Maryland, Virginia, Pennsylvania, West Virginia, and the District of Columbia.

Midwest Clean Diesel Initiative

www.epa.gov/midwestcleandiesel

EPA Region 5's Midwest Clean Diesel Initiative is a cooperative, public–private effort to reduce diesel emissions along major transportation corridors from various sectors including trucking, locomotive, construction, and ports, with an emphasis on urban areas in the Midwest.

Southeast Diesel Collaborative

www.southeastdiesel.org

The Southeast Diesel Collaborative is a voluntary, public–private partnership involving leaders from federal, state, and local governments; the private sector; and other stakeholders throughout the Southeast working to reduce diesel emissions.

Rocky Mountain Clean Diesel Collaborative

www.epa.gov/region8/air/rmcdc.html

The Rocky Mountain Clean Diesel Collaborative is a partnership of federal, state, and local governments; nonprofit organizations; the private sector; and environmental groups in Colorado, Montana, North Dakota, South

Dakota, Utah, and Wyoming. The collaborative was established in November 2006.

Blue Skyways Collaborative

www.blueskyways.org

The Blue Skyways Collaborative was created to encourage voluntary air emission reduction in North America's heartland. Through partnership with nonprofit and environmental groups; private industries; and international, federal, state, and local governments, Blue Skyways strives to improve air quality.

West Coast Diesel Collaborative

www.westcoastcollaborative.org

EPA Regions 9 and 10 were instrumental in creating the West Coast Diesel Collaborative. This initiative is a partnership among leaders from federal, state, and local government; the private sector; and environmental groups in California, Oregon, Idaho, Washington, Alaska, Canada, and Mexico, committed to reducing diesel emissions along the West Coast.

Acronyms and Abbreviations

APU	Auxiliary Power Unit
CARB	California Air Resources Board
CCV	Closed Crankcase Ventilation
CFS	Crankcase Filtration System
CNG	Compressed Natural Gas
CO	Carbon Monoxide
CO_2	Carbon Dioxide
DEQ	Diesel Emissions Quantifier
DERA	Diesel Emissions Reduction Act
DOC	Diesel Oxidation Catalyst
DPF	Diesel Particulate Filter
EGR	Exhaust Gas Recirculation

EPA	U.S. Environmental Protection Agency
EPAct	Energy Policy Act
FY	Fiscal Year
HC	Hydrocarbon
LNT	Lean Nitrogen Oxides Trap
NAAQS	National Ambient Air Quality Standards
NCDC	National Clean Diesel Campaign
NESCAUM	Northeast States for Coordinated Air Use Management
NMHC	Nonmethane Hydrocarbon
NO_2	Nitrogen Dioxide
NO_X	Nitrogen Oxides
$PM_{2.5}$	Particulate Matter (Fine)
PPM	Parts Per Million
RFP	Request for Proposal
RIA	Regulatory Impact Analysis
SCR	Selective Catalytic Reduction
SCRT	Selective Catalytic Reduction Technology
SIP	State Implementation Plan
SO_X	Sulfur Oxides
ULSD	Ultra-Low-Sulfur Diesel

End Notes

[1] Congress also provided $10 million for grants for cost-effective emission reduction projects for two California air quality management districts (South Coast and San Joaquin).

[2] *National Ambient Air Quality Standards for Particle Pollution Regulatory Impact Assessment* (PM NAAQS RIA 10-06-06, Chapter 3, pg. 3-14, Table 3.2): Compares other stationary and area sources for PM reduction. Report is available at www.epa.gov/ttn/ecas/ria.html and www.epa.gov/ttn/ecas/regdata/RIAs/Chapter%203--Controls.pdf.

[3] *The Cost-Effectiveness of Heavy-Duty Diesel Retrofits and Other Mobile Source Emissions Reduction Projects and Programs* (EPA-420-B-07-006, May 2007, Appendix Table 3, pg. 13-15): Compares other mobile source programs cost-effectiveness for NO_x/VOCs. Report is available at www.epa.gov/otaq/stateresources/policy/general/420b07006.pdf.

[4] The weighted average federal cost of DERA projects is about $25,000 per ton PM and $1,500 per ton NO_x.

[5] As of December 16, 2008, there are 57 eight-hour ozone nonattainment areas (1997) that consist of 293 full or partial counties, with a total population of almost 132 million. An additional 74 counties, where 16 million people reside, show air quality values that do not meet the 2008 ozone standard. As of December 16, 2008, there are 39 $PM_{2.5}$ nonattainment areas (1997) that consist of 208 full or partial counties, with a total population exceeding 88 million. On December 22, 2008, EPA designated nonattainment areas for the 2006 $PM_{2.5}$ NAAQS. As of December 22, 2008, there are 58 $PM_{2.5}$ nonattainment areas (2006) that consist of 211 full or partial counties.

[6] Many of the benefits will accrue in the first five years of the program, and the benefits are discounted using a 3 percent rate.

[7] The District of Columbia is now an eligible entity for the State Clean Diesel Grant and Loan Program, but it was not eligible when EPA awarded the FY 2008 state grants.

[8] The District of Columbia is now an eligible entity for the State Clean Diesel Grant and Loan program, but it was not eligible when EPA awarded the FY 2008 state grants.

[9] *The Cost-Effectiveness of Heavy-Duty Diesel Retrofits and Other Mobile Source Emission Reduction Projects and Programs* (EPA-420-B-07-006, May 2007)—Tables 3 and 4 provide comparisons of cost-effectiveness of other mobile source programs; and *Diesel Retrofit Technology: An Analysis of the Cost-Effectiveness of Reducing Particulate Matter and Nitrogen Oxides Emissions From Heavy-Duty Nonroad Diesel Engines Through Retrofits* (EPA-420-R-07-005, May 2007). The full reports are available on the EPA Web site: www.epa.gov/otaq/diesel/publications.htm.

[10] The estimates are based on the peer-reviewed studies on the relationship between $PM_{2.5}$ and premature mortality from:

Pope, C.A., III, R.T. Burnett, M.J. Thun, E.E. Calle, D. Krewski, K. Ito, and G.D. Thurston. 2002. Lung Cancer, Cardiopulmonary Mortality, and Long-Term Exposure to Fine Particulate Air Pollution. *Journal of the American Medical Association*. 287: 1132-1141.

Laden, F., J. Schwartz, F.E. Speizer, and D.W. Dockery. 2006. Reduction in Fine Particulate Air Pollution and Mortality. *American Journal of Respiratory and Critical Care Medicine*. 173: 667-672.

For more information about this approach, please see EPA's 2008 *Technical Support Document: Calculating Benefit Per-Ton Estimates*, Ozone NAAQS Docket #EPA-HQ-OAR-2007-0225-0284. The premature mortality estimates are generated using $PM_{2.5}$ co-benefits data from EPA's 2006 *Ozone Final Regulatory Impact Analysis (RIA)*, NAAQS for Particulate Matter, Chapter 5 Benefit Analysis and Results. These techniques are described in a memo to the file: Memo From Michael Wolfe, Ken Davidson, Rosalva Tapia, Kuang Wei, and Jen Went to Jim Blubaugh, Manager, Innovative Strategies Group, EPA Office of Transportation and Air Quality. March 25, 2009. Energy Policy Act of 2005 Diesel Emissions Reduction (DERA) Program FY 2008: Benefit and Cost-Effectiveness Calculation Methodologies.

In: Diesel Emissions: Reduction Efforts ISBN: 978-1-62081-561-8
Editors: J. F. Price and S. E. Mitchell

Chapter 2

DIESEL POLLUTION: FRAGMENTED FEDERAL PROGRAMS THAT REDUCE MOBILE SOURCE EMISSIONS COULD BE IMPROVED*

United States Government Accountability Office

WHY GAO DID THIS STUDY

Exhaust from diesel engines is a harmful form of air pollution. EPA has issued emissions standards for new diesel engines and vehicles, but older mobile sources of diesel emissions—such as trucks and buses—continue to emit harmful pollution. Programs at DOE, DOT, and EPA provide funding for activities that reduce diesel emissions, such as retrofitting existing diesel engines and vehicles. The existence of these programs at multiple agencies has raised questions about the potential for unnecessary duplication. In response to a mandate in the Diesel Emissions Reduction Act of 2010, GAO examined the (1) extent of duplication, overlap, fragmentation, or gaps, if any, among federal grant, rebate, and loan programs that address mobile source diesel emissions; (2) effectiveness of federal funding for activities that reduce mobile source diesel emissions; and (3) extent of collaboration among agencies that

* This is an edited, reformatted and augmented version of the Highlights of GAO-12-261, a report to congressional committees, dated February 2012.

fund these activities. GAO analyzed program data, documents, and relevant laws and regulations and interviewed agency officials. GAO also reviewed three diesel-related tax expenditures.

What GAO Recommends

GAO recommends that DOT's Federal Transit Administration develop performance measures for its two relevant strategic goals and that DOE, DOT, and EPA establish a strategy for collaboration among their programs that fund activities that reduce diesel emissions. DOE and EPA agreed with the relevant recommendation, and DOE questioned several findings. DOT questioned several findings and both recommendations and neither agreed nor disagreed with the recommendations. GAO continues to believe in the need for the performance measures and collaboration.

What GAO Found

Federal grant and loan funding for activities that reduce mobile source diesel emissions is fragmented across 14 programs at the Department of Energy (DOE), the Department of Transportation (DOT), and the Environmental Protection Agency (EPA). From fiscal years 2007 through 2011, the programs obligated at least $1.4 billion for activities that have the effect of reducing mobile source diesel emissions. The programs have varying goals and purposes; nevertheless, each program allows or requires a portion of its funding to support activities that reduce mobile source diesel emissions, such as replacing fleets of older diesel trucks or school buses with natural gas vehicles. In addition, each of the 14 programs overlaps with at least one other program in the specific activities they fund, the program goals, or the eligible recipients of funding. GAO also identified several instances of duplication where more than one program provided grant funding to the same recipient for the same type of activities. However, GAO was unable to determine whether unnecessary duplication exists because of limited information on program administrative costs, among other things. GAO did not find any gaps among the programs, such as mobile sources that are not eligible for funding.

The effectiveness of federal funding for activities that reduce mobile source diesel emissions is unknown because agencies vary in the extent to which they have established performance measures. DOE and EPA have established performance measures for the strategic goals related to their programs that reduce mobile source diesel emissions.

DOT has established such measures for two of its administrations—the Federal Aviation Administration and Federal Highway Administration—but has not established such measures for the Federal Transit Administration for two of the four strategic goals that link to its programs that fund diesel emissions reduction activities. Instead, agency officials said they collect information on the current condition of the nation's transit fleet, among other things, to measure the performance of its programs.

As GAO has previously reported, principles of good governance indicate that agencies should establish quantifiable performance measures to demonstrate how they intend to achieve their goals and measure the extent to which they have done so. In addition, 13 of the 14 programs have purposes other than decreasing diesel emissions, and diesel reductions are a side benefit of efforts to achieve these other goals.

As a result, few programs collect diesel-related performance information. Incomplete performance information may limit the ability of agencies to assess the effectiveness of their programs and activities that reduce diesel emissions.

The programs that fund activities that reduce diesel emissions generally do not collaborate because of the differing purposes and goals of each program, according to senior DOE, DOT, and EPA officials.

The officials also were sometimes unaware of other programs that fund similar activities and said that any existing collaboration was on a case-by-case basis. GAO's previous work has shown that although federal programs have been designed for different purposes, coordination among programs with related responsibilities is essential to efficiently and effectively meet national concerns. Further, without a coordinated approach, programs can waste scarce funds, confuse and frustrate program customers, and limit the overall effectiveness of the federal effort.

ABBREVIATIONS

DOE	Department of Energy
DOT	Department of Transportation
EPA	Environmental Protection Agency

February 7, 2012

The Honorable Barbara Boxer
Chairman
The Honorable James M. Inhofe
Ranking Member
Committee on Environment and Public Works
United States Senate

The Honorable Fred Upton
Chairman
The Honorable Henry A. Waxman
Ranking Member
Committee on Energy and Commerce
House of Representatives

Diesel engines play a vital role in public transportation, construction, agriculture, and shipping, largely because they are more durable and reliable than gasoline-powered engines, as well as 25 to 35 percent more energy efficient. However, exhaust from diesel engines is a more pervasive and harmful form of air pollution than exhaust from gasoline-powered engines. Diesel exhaust contains air pollutants such as nitrogen oxides and particulate matter, as well as other harmful substances that affect public health and the environment.[1] Since 1984, the Environmental Protection Agency (EPA) has implemented standards that have progressively lowered the maximum allowable amount of certain pollutants, including nitrogen oxides and particulate matter, from new diesel engines by more than 98 percent. However, the most stringent standards generally apply to diesel engines and vehicles built after 2007, and EPA estimates that more than 20 million mobile sources of diesel emissions built before 2007—13 million on-highway vehicles, 7 million nonroad engines, and 47,000 locomotive and marine engines—continue to emit higher amounts of harmful pollutants than newer engines.[2]

Programs at the Department of Energy (DOE), the Department of Transportation (DOT), and EPA address mobile source diesel emissions by funding projects that, among other things, retrofit, rebuild, or replace existing diesel engines or vehicles; install devices that reduce idling of diesel engines; and convert diesel engines and vehicles to use cleaner fuels, such as natural gas or propane. The existence of these programs at multiple agencies has raised questions about the potential for unnecessary duplication. We have previously reported that fragmentation and overlap among government programs can lead to such duplication.[3] Fragmentation occurs when more than one federal agency, or more than one organization within an agency, is involved in the same broad area of national need. Overlap occurs when multiple agencies and programs have similar goals, engage in similar activities or strategies to achieve them, or target similar beneficiaries. We have also reported that federal programs contributing to the same or similar outcomes should closely coordinate to improve their overall effectiveness.[4]

In response to a mandate in the Diesel Emissions Reduction Act of 2010, this report examines the (1) extent to which duplication, overlap, fragmentation, or gaps, if any, exist among federal grant, rebate, and loan programs that address mobile source diesel emissions; (2) effectiveness of federal funding for activities that reduce mobile source diesel emissions; and (3) extent to which collaboration takes place among agencies that fund activities that reduce mobile source diesel emissions.[5] This report also provides information on tax expenditures that address mobile source diesel emissions.[6]

To address the first objective, we reviewed relevant statutes and regulations; conducted a literature review; analyzed agency documents; interviewed agency officials and industry stakeholders; and conducted Internet searches to identify grant, rebate, and loan programs that fund mobile source diesel emissions reduction activities. We then reviewed agency documents and conducted structured interviews of agency officials from each relevant DOE, DOT, and EPA program we identified. We conducted interviews about the program's purpose, goals, eligible activities, target beneficiaries, and types of funding to identify areas of duplication, overlap, or fragmentation. We also compared the types of eligible recipients under each program with available data on the sources of diesel emissions to identify any gaps among the programs. In addition, we obtained and analyzed funding data from the three agencies to estimate the total amount of federal funding for diesel emissions reduction projects from fiscal years 2007 through 2011. We selected this period because, by 2007, EPA had issued emissions standards for key on-road sources, such as heavy-duty trucks and buses, as well as a rule requiring refiners to reduce the sulfur content of—and therefore emissions from—certain diesel fuels. We reviewed documents about the underlying databases and interviewed knowledgeable agency officials to assess the reliability of the data for each program. We determined that the data obtained from these agencies were sufficiently reliable for the purposes of this report. To respond to the second and third objectives, we reviewed agency documents and conducted structured interviews with agency officials about program goals and performance, coordination with other programs, and assessments of diesel pollution, among other things. We then compared these programs' efforts with best practices for federal programs that contribute to the same outcome. To identify these best practices, we reviewed our prior work as well as relevant statutes, including the Government Performance and Results Act of 1993 and the Government Performance and Results Act Modernization Act of 2010.[7] A more detailed description of our scope and methodology is presented in appendix I.

We conducted this performance audit from May 2011 to February 2012 in accordance with generally accepted government auditing standards. Those standards require that we plan and perform the audit to obtain sufficient, appropriate evidence to provide a reasonable basis for our findings and conclusions based on our audit objectives. We believe that the evidence obtained provides a reasonable basis for our findings and conclusions based on our audit objectives.

BACKGROUND

Under the Clean Air Act, EPA is responsible for setting National Ambient Air Quality Standards for certain pollutants considered harmful to public health and the environment. EPA has set these standards for six such pollutants, known as criteria air pollutants: carbon monoxide, nitrogen oxides, sulfur oxides, particulate matter, ozone, and lead.[8] Diesel exhaust contains nitrogen oxides, particulate matter, and numerous other harmful chemicals. Exposure to nitrogen oxides can result in adverse respiratory effects, and nitrogen oxides contribute to the formation of ozone, which can cause respiratory illnesses, decreased lung function, and premature death. A large body of scientific evidence links exposure to particulate matter to serious health problems, including asthma, chronic bronchitis, heart attack, and premature death.

According to EPA documents, as of 2009, mobile diesel sources emitted about 47 percent (6.4 million tons) of the nation's nitrogen oxides and about 16 percent (300,000 tons) of its particulate matter.[9] EPA estimated that on-highway trucks and vehicles and nonroad sources, such as agriculture and construction equipment, contributed about 57 percent of these nitrogen oxide emissions and 58 percent of the particulate matter emissions, and marine and locomotive sources contributed the remaining 43 percent and 42 percent of such emissions, respectively, from diesel engines.

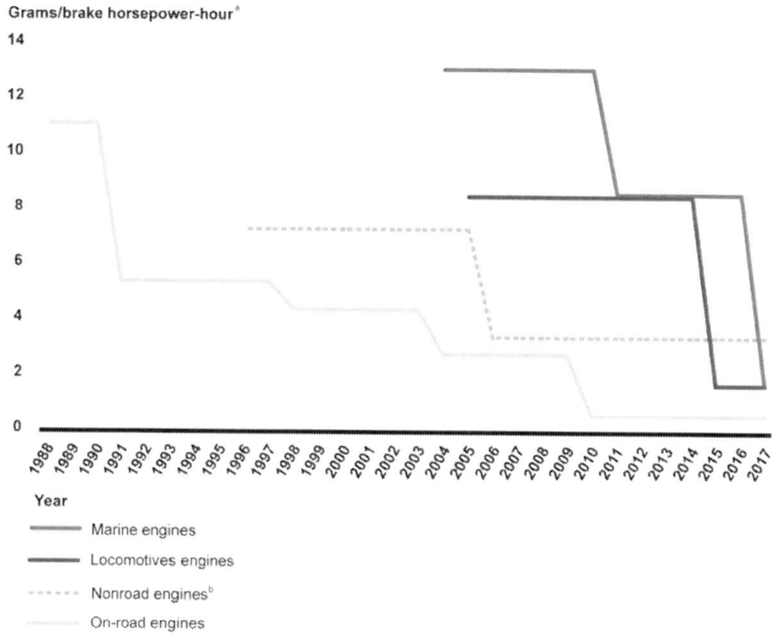

Source: GAO analysis.

[a] Grams per brake horsepower-hour is a measure of the grams of nitrogen oxides a vehicle or engine emits per the amount of energy the vehicle or engine uses during one hour.

[b] These emission limits apply to nonroad engines at or above 175 horsepower, such as a large bulldozer engine. EPA has also set nitrogen oxide emission standards for such engines below 50 horsepower, such as small tractor engines, and engines between 50 and 175 horsepower, which took effect in 2006 and 2008, respectively.

Figure 1. Effective Dates of Major Nitrogen Oxide Emission Limits for Newly Manufactured Mobile Sources of Diesel Emissions, by Source.

EPA has progressively implemented more stringent diesel emissions standards to lower the amount of key pollutants from mobile diesel sources since 1984.[10] For example, EPA regulations for heavy-duty highway diesel engines required a 98 percent reduction from 1988 allowable levels of nitrogen oxide and particulate matter emissions for new engines built after 2009. The most recent emissions standards for construction and agricultural equipment began to take effect in 2008 and required a 95 percent reduction in nitrogen oxides and a 90 percent reduction in particulate matter from previous standards, which took effect in 2006 and 2007. In 2008, EPA issued its most recent regulations for new marine vessels and locomotives, which EPA expects will, by 2030, reduce nitrogen oxide emissions from the engines of these sources by about 80 percent and particulate matter emissions by about 90 percent compared to previous standards. Figures 1 and 2 show the effective dates of major reductions in allowable amounts of nitrogen oxide and particulate matter emissions from mobile diesel sources.

Owners and operators of diesel engines can undertake a variety of activities to reduce diesel emissions, including retrofitting, rebuilding, or replacing existing diesel engines or vehicles; installing devices that reduce idling of diesel engines; and converting diesel engines and vehicles to use cleaner fuels. Retrofitting existing diesel engines generally involves the installation of emissions control devices, such as filters, on a vehicle's tailpipe. Rebuilding components of existing diesel engines can return engines to their original emissions levels or involve the installation of new technology that produces lower levels of emissions. Replacing existing diesel engines and vehicles with newer, lower emitting engines or vehicles can lead to significant emissions reductions, but because it is a costly option, it may be most appropriate for the oldest, most polluting vehicles. Devices that reduce idling of diesel engines generally allow a vehicle's heat, air conditioning, and other electrical equipment to run without operation of the vehicle's main engine. Converting diesel vehicles and engines to use cleaner fuels can also provide significant emissions reductions.

The Government Performance and Results Act, as amended, requires agencies to prepare annual performance plans that contain, among other things, a set of annual goals that establish the agencies' intended performance and measures that can be used to assess progress toward achieving those goals. DOE, DOT, and EPA establish and organize these goals and performance measures at differing agency and administrative levels. Specifically, DOE and EPA establish strategic goals and performance measures for each goal as part of their agencywide performance plans. DOT establishes strategic goals as part of its agencywide strategic plan, but the agency's administrations—the Federal Aviation Administration, Federal Highway Administration, and Federal Transit Administration, among others—generally establish their own performance measures for assessing their programs' contributions to the department's strategic goals.

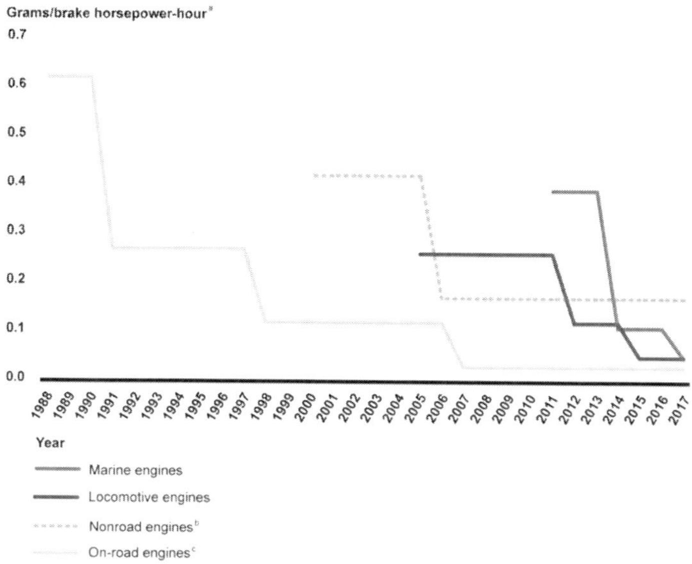

Source: GAO analysis

[a] Grams per brake horsepower-hour is a measure of the grams of particulate matter a vehicle or engine emits per the amount of energy the vehicle or engine uses during one hour.

[b] These emission limits apply to nonroad engines at or above 175 horsepower, such as a large bulldozer engine. EPA has also set particulate matter emission standards for such nonroad engines below 50 horsepower, such as small tractor engines, which took effect in 2000 and 2006.

[c] The 1998 emission limit in the figure applies to urban buses. EPA also set particulate matter standards for heavy-duty trucks that took effect in 1998, which limited emissions from these engines to 0.1 grams per brake horsepower-hour.

Figure 2. Effective Dates of Major Particulate Matter Emission Limits for Newly Manufactured Mobile Sources of Diesel Emissions, by Source.

Federal Funding for Activities That Reduce Diesel Emissions Is Fragmented across 14 Programs

Federal grant and loan funding for activities that reduce mobile source diesel emissions is fragmented across 14 programs at DOE, DOT, and EPA. Many of these programs generally target air pollution, but of the 14 programs, one—EPA's Diesel Emissions Reduction Act program—has a specific purpose of reducing mobile source diesel emissions.

The remaining 13 programs focus on other goals or purposes, such as supporting energy efficiency projects or reducing petroleum use. Nevertheless, each of these programs allows or requires a portion of its funding to support activities that have the effect of reducing mobile source diesel emissions. For example, authorizing legislation for DOT's Congestion Mitigation and Air Quality Improvement program directs grant recipients to give priority to certain activities, including retrofitting diesel engines and vehicles.

The 14 programs provide funding through one or more mechanisms, including competitive grants, formula grants, and loans.[11] Specifically, 13 of the programs provide funding through competitive and formula grants, and 1 program—DOT's State Infrastructure Banks program—provides loans.[12]

We did not identify any gaps in the programs, such as mobile sources that are not eligible for funding. See appendix II for additional information about each program.

From fiscal years 2007 through 2011, these 14 programs obligated at least \$1.4 billion for activities that have the effect of reducing mobile source diesel emissions.[13]

According to data from DOE, DOT, and EPA, the American Recovery and Reinvestment Act of 2009 provided about \$870 million of this funding. The \$1.4 billion amount is a lower bound because DOT could not determine the amount of grant and loan funding some of its programs have provided for projects that reduce mobile source diesel emissions. According to DOT officials, the agency does not track this information because statutory program requirements do not call for the agency to do so.[14]

These activities that have the effect of reducing mobile source diesel emissions include replacing fleets of older diesel trucks or school buses with natural gas vehicles, installing particulate matter filters on construction equipment, and replacing diesel-powered airport luggage transporters with electric vehicles.

As table 1 shows, some of the programs that support these activities have broad purposes, such as increasing energy efficiency in transportation, reducing petroleum consumption, or funding public transportation projects, and other programs have narrower purposes, such as reducing emissions at airports, constructing ferry boats and related facilities, or promoting alternative transportation systems in and around national parks.

Table 1. Estimated Federal Funds Obligated for Mobile Source Diesel Emissions Reduction Activities, by Agency and Program, Fiscal Years 2007-2011

Dollars in millions			
Agency/program	**Purpose**	**Grants**	**Loans**
DOE			
Clean Cities program	To advance the nation's economic, environmental, and energy security by funding projects that reduce petroleum use in transportation	$305	—
Energy Efficiency and Conservation Block Grant program	To support energy efficiency and conservation projects that reduce fossil fuel emissions and energy use and improve energy efficiency in the transportation and building sectors	256	—
State Energy Program	To support state development and implementation of strategies and goals that promote energy efficiency and conservation	11	—
DOE total		**$572**	**—**
DOT			
Federal Aviation Administration			
Voluntary Airport Low Emissions program	To reduce airport ground vehicle and equipment emissions at airports in air quality nonattainment and maintenance areas	18	—
Federal Highway Administration			
Congestion Mitigation and Air Quality Improvement program	To support transportation projects that contribute to the attainment or maintenance of carbon monoxide, ozone, and particulate matter air quality standards and relieve congestion	178	—
Ferry Boat Discretionary program	To fund the construction of ferry boats and ferry terminal facilities	26	—
State Infrastructure Banks program	To facilitate state establishment of infrastructure banks to provide nongrant assistance for eligible transportation projects	—	$6[a]
Federal Transit Administration			
Bus and Bus Facilities program	To replace, rehabilitate, and purchase buses and related equipment and construct or rehabilitate bus-related facilities	—[b]	—
Clean Fuels Grant program	To assist achievement and maintenance of air quality standards by providing grants for clean fuel buses and facilities	17	—
National Fuel Cell Bus Technology Development program	To develop commercially viable fuel cell bus technology and related infrastructure	11	—
Transit in Parks program	To promote alternative transportation systems in and around national parks and other federal lands	—[c]	—
Transit Investments for Greenhouse Gas and Energy Reduction program	To reduce energy consumption or greenhouse gas emissions of public transportation systems	96	—

Table 1. (Continued)

Dollars in millions			
Agency/program	**Purpose**	**Grants**	**Loans**
Urbanized Area Formula Grants program	To support capital projects, planning, transit enhancements, and certain operating costs for public transportation in urbanized areas	—[b]	—
DOT total		**$346[d]**	**$6**
EPA			
Diesel Emissions Reduction Act Program	To reduce emissions from existing diesel engines and vehicles	512	—
Estimated total federal funding		**$1,430[d]**	**$6**

Source: GAO analysis of relevant laws and DOE, DOT, and EPA data and documents.

Notes: The American Recovery and Reinvestment Act of 2009 provided funding for DOE's Clean Cities, Energy Efficiency and Conservation Block Grant, and State Energy programs; DOT's Ferry Boat Discretionary, Transit Investments in Greenhouse Gas and Energy Reduction, and Urbanized Area Formula Grants programs; and EPA's Diesel Emissions Reduction Act Program.

[a] This amount is a lower bound because Federal Highway Administration officials do not obtain complete data from states on projects funded through the State Infrastructure Bank program.

[b] DOT was unable to provide information on the amount of funding this program awarded for projects that reduced mobile source diesel emissions.

[c] We were unable to identify any projects this program funded during fiscal years 2007-2011 that reduced mobile source diesel emissions. However, this program does allow its funding to support activities that reduce these emissions.

[d] This total is a lower bound because agencies were unable to determine the amount of funding that some of their programs provided for activities that reduce diesel emissions.

As table 2 shows, each of the 14 relevant programs overlaps with at least one other program in the specific types of activities they fund, the program goals, or the eligible recipients of funding. For example, 6 of the 14 programs share a broad goal of increasing energy efficiency, and local governments are eligible to receive grants under 10 of the programs. In addition, we found that 13 of the 14 programs fund activities that retrofit diesel engines or vehicles, and 11 programs fund activities that reduce diesel vehicle idling. We also identified the potential for overlap among these 11 programs and an excise tax exemption for certain vehicle idling reduction devices because the tax expenditure and the 11 programs all provide incentives to use idle reduction devices to reduce diesel emissions.

Table 2. Overlapping Mobile Source Diesel Emissions Reduction Activities, Goals, and Eligible Recipients, by Agency and Program

Agency/program	Activities					Goals				Eligible recipients					
	Retrofit vehicle or engine	Rebuild vehicle or engine	Replace vehicle or engine	Reduce vehicle idling	Use cleaner fuel	Reduce emissions	Reduce pollution in areas not meeting air quality standards	Increase energy efficiency	Reduce fuel use	State governments	Local governments	Land management agencies	Transit agencies	Federally recognized tribes	Private or nonprofit organizations
DOE															
Clean Cities	•		•	•	•				•	•	•	•	•	•	•
Energy Efficiency and Conservation Block Grant	•	•	•	•	•	•		•		•	•			•	
State Energy Program	•	•	•	•	•			•	•	•					
DOT															
Federal Aviation Administration															
Voluntary Airport Low Emissions	•		•		•	•	•			•	•				
Federal Highway Administration															
Congestion Mitigation and Air Quality Improvement	•	•	•	•	•		•			•	•		•		
Ferry Boat Discretionary	•	•	•		•					•	•				•

Table 2. (Continued)

Agency/program	Activities					Goals				Eligible recipients					
	Retrofit vehicle or engine	Rebuild vehicle or engine	Replace vehicle or engine	Reduce vehicle idling	Use cleaner fuel	Reduce emissions	Reduce pollution in areas not meeting air quality standards	Increase energy efficiency	Reduce fuel use	State governments	Local governments	Land management agencies	Transit agencies	Federally recognized tribes	Private or nonprofit organizations
State Infrastructure Banks	●	●	●	●	●					●					
Federal Transit Administration															
Bus and Bus Facilities	●	●	●		●				●	●	●		●	●	
Clean Fuels Grant	●	●	●	●	●	●	●	●	●	●	●		●		
National Fuel Cell Bus Technology Development			●	●	●	●		●	●						●
Transit in Parks	●	●	●	●	●	●				●	●	●	●	●	●
Transit Investments in Greenhouse Gas and Energy Reduction	●	●	●	●	●	●		●	●	●			●	●	
Urbanized Area Formula Grants	●	●	●	●	●				●	●	●		●		
EPA															
Diesel Emissions Reduction Act Program	●	●	●	●	●	●				●	●		●	●	●

Source: GAO analysis of relevant laws and DOE, DOT, and EPA documents and interviews.

Appendix III provides additional information on this and two other tax expenditures related to diesel emissions reductions.

We also identified several instances of duplication where more than one program provided funding to the same recipient for the same type of activities. In one case, a state transportation agency received \$5.4 million from DOT's Transit Investments in Greenhouse Gas Emissions Reduction program to, among other things, upgrade 37 diesel buses to hybrid diesel-electric buses; \$3.5 million from DOT's Congestion Mitigation and Air Quality Improvement program to replace diesel buses with 4 hybrid diesel-electric buses; and \$2.3 million from DOT's Clean Fuels Grant program to replace 4 diesel buses with hybrid electric buses. In another case, a nonprofit organization received \$1.1 million from EPA's Diesel Emissions Reduction Act Program to install emission reduction and idle reduction technologies on 1,700 trucks as well as \$5.6 million from a state infrastructure bank established under DOT's program to equip trucks and truck fleets with emissions control and idle reduction devices.

Even with duplication among the programs, several factors make it difficult to precisely determine whether unnecessary duplication exists. First, when different programs fund the same diesel emissions reduction activities, it is not necessarily wasteful. For example, a transit agency could use funds from two different programs to replace two separate fleets of aging diesel buses. Second, grant recipients may leverage funding from more than one program to support the full cost of diesel emissions reduction projects. In some cases, grant recipients have used funding from multiple agencies, in addition to local matching funds, to support the cost of large projects that include multiple diesel emissions reduction activities. Third, agencies were often unable to provide information necessary to determine whether and to what extent unnecessary duplication exists among the programs. For example, several agencies reported that they do not track costs for administrative functions at the program level. Without information on these costs, it is difficult to determine whether and to what extent programs perform duplicative administrative functions that could be consolidated to provide grants and loans more efficiently.

The fragmentation, overlap, and duplication among these programs result, in part, from their legislative creation as separate programs with different purposes that fund a wide range of activities, some of which have the effect of reducing mobile source diesel emissions. We have previously reported that, as the federal government has responded over time to new needs and problems, many agencies have been given responsibilities for addressing the same or similar national issues.[15] Some of this shared responsibility was intended to recognize that addressing some issues from a national perspective would necessarily involve more than one agency or approach. However, the resulting fragmentation, overlap, and duplication may waste administrative resources and create an environment in which participants are not served as efficiently and effectively as possible. In addition, we have previously reported that fragmentation, overlap, and duplication suggest the need for further examination of programs to identify potential areas for improvement, realignment, consolidation, or elimination.[16]

THE EFFECTIVENESS OF FEDERAL FUNDING FOR DIESEL EMISSIONS REDUCTION ACTIVITIES IS UNKNOWN

The effectiveness of federal funding for activities that reduce mobile source diesel emissions is unknown because agencies vary in the extent to which they have established performance measures. In addition, few programs collect performance information on their diesel emissions reduction activities because 13 of the 14 programs that fund these activities have purposes other than reducing diesel emissions. This incomplete performance information may limit the ability of agencies to assess the effectiveness of their programs and activities that reduce diesel emissions.

Agencies that fund activities that reduce mobile source diesel emissions have established performance measures for their strategic goals to varying degrees. DOE and EPA have established performance measures for the strategic goals related to their programs that reduce mobile source diesel emissions. For example, EPA monitors progress toward its strategic goal of reducing greenhouse gas emissions and developing adaptation strategies to protect and improve air quality by measuring, among other things, the tons of mobile source emissions its programs reduce.

DOT has established such performance measures for two of its administrations—the Federal Aviation Administration and Federal Highway Administration—but has not established such measures for the Federal Transit Administration for two of the four strategic goals that link to its programs that fund diesel emissions reduction activities.

Appendix IV provides additional information on these agencies' strategic goals and performance measures related to programs that reduce mobile source diesel emissions.

The Government Performance and Results Act, as amended, generally requires agencies to provide a basis for comparing actual results with established goals, and as such, federal departments and agencies are to comply with Government Performance and Results Act requirements.

As we have previously reported, Government Performance and Results Act requirements also can serve as leading practices at lower levels within federal agencies, such as individual divisions, programs, or initiatives.[17]

We have also reported that principles of good governance indicate that agencies should establish quantifiable performance measures to demonstrate how they intend to achieve their goals and measure the extent to which they have done so.[18] The Federal Transit Administration has not established performance measures for its goals of (1) environmental sustainability—that is, advancing environmentally sustainable policies and investments that reduce carbon and other harmful emissions from transportation sources—and (2) economic competitiveness—that is, promoting transportation policies and investments that bring lasting and equitable economic benefits to the nation and its citizens.

Agency officials said they generally collect information on the current condition of the nation's transit fleet, the use of public transportation, and transit fleet compliance with the Americans with Disabilities Act to measure the performance of the agency's transit programs.[19] However, this information will not enable the agency to determine the extent to which it has met its goals related to environmental sustainability and economic competitiveness.

At the program level, limited performance information is available about the results of activities that reduce mobile source diesel emissions. The 14 programs that fund activities that reduce diesel emissions currently collect performance information on their diesel emissions reduction activities to varying degrees.

According to agency documents and officials, EPA's Diesel Emissions Reduction Act Program collects performance information on the amount and type of diesel emissions reductions each project achieves; DOE's three programs collect some emissions reduction information but do not quantitatively collect diesel emissions reduction information; three of DOT's programs collect some performance information related to diesel emissions reductions; and the remaining seven DOT programs do not collect performance information related to diesel emissions.

This variation in the amount of diesel-related performance information programs collect occurs partially because 13 of the 14 programs that fund these activities have purposes other than reducing diesel emissions, such as supporting energy efficiency projects or reducing petroleum use. However, without information on the results of the programs' activities that reduce mobile source diesel emissions, the overall effectiveness of federal grant and loan funding for activities that reduce diesel emissions cannot be determined.

EPA

EPA's Diesel Emissions Reduction Act Program collects information on the number of diesel engines it replaces, retrofits, and rebuilds as well as information on the estimated tons of particulate matter, nitrogen oxide, carbon dioxide, carbon monoxide, and hydrocarbon emissions it reduces. According to agency documents, in fiscal year 2008—the most recent year for which data were reported—the program reduced approximately 46,000 tons of nitrogen oxide emissions and 2,200 tons of particulate matter emissions.[20] EPA documents show that the cost for these emissions reductions ranged from $400 to $2,000 per ton of nitrogen oxide emissions reduced and from $9,000 to $27,700 per ton of particulate matter emissions reduced.

DOE

DOE's Clean Cities program collects information on reductions in gasoline and diesel fuel use that the program achieves to measure progress toward its program goal of reducing national petroleum use by 2.5 billion gallons by 2020. DOE's Energy Efficiency and Conservation Block Grant and State Energy programs estimate emissions reductions that result from program activities, but neither of these programs separately tracks diesel emissions from other emissions reductions.

DOT's Federal Aviation Administration

The Federal Aviation Administration's Voluntary Airport Low Emissions program collects information on the total amount of criteria pollutant emissions each project will reduce, but it does not currently track reductions in diesel emissions.

DOT's Federal Highway Administration

The Federal Highway Administration's Congestion Mitigation and Air Quality Improvement program collects information from grant recipients on the type and quantity of emissions reduced through each project the program funds. However, the program does not review or compile this information at the national level. The Ferry Boat Discretionary and State Infrastructure Banks programs do not collect performance information related to diesel emissions reductions.

DOT's Federal Transit Administration

The Federal Transit Administration's Transit Investments for Greenhouse Gas and Energy Reduction program obtains information from grant applicants on the amount of energy use and greenhouse gas emissions each project is to reduce, but the program does not separately track reductions in diesel energy use or diesel emissions. The remaining five Federal Transit Administration programs that fund diesel emissions reduction activities—Bus and Bus Facilities, Clean Fuels Grant, National Fuel Cell Bus Technology Development, Transit in Parks, and Urbanized Area Formula Grants—do not collect performance information related to diesel emissions reductions.

Efforts to measure the effects of programs that decrease diesel emissions are also hindered by the absence of a baseline assessment of nationwide diesel emissions from which agencies could measure progress. EPA has assessed national levels of nitrogen oxide and particulate matter pollution from some mobile diesel sources, including highway vehicles and some nonroad equipment, and DOT maintains data on the number of diesel transit vehicles currently in use. However, no agency has comprehensively assessed existing diesel pollution to identify the most significant mobile sources of diesel emissions and the specific areas that face the greatest health risks from diesel pollution. Without a more comprehensive assessment, agencies cannot identify and target, within their discretion, funding toward specific sectors or geographic areas of greatest need.[21] Agencies generally provide funds to recipients based on criteria that may derive from law, agency discretion, or a combination thereof. Under some programs, agencies allocate funding based on statutory formulas or criteria. For example, DOT's Urbanized Area Formula Grants program uses a statutory formula to allocate funds on the basis of population and population density. EPA's Diesel Emissions Reduction Act program awards funds competitively but, based on statutory criteria, must prioritize projects that maximize health benefits, are the most cost-effective, and serve areas with poor air quality, among other factors. Under other programs, agencies have some discretion in awarding funds. These agencies generally consider applicant eligibility and other relevant factors, but this does not include consideration of which areas face the greatest diesel-related health risks.

Limited Collaboration May Hinder the Effectiveness of Federal Funding for Activities That Reduce Diesel Emissions

The federal programs that fund activities that have the effect of reducing mobile source diesel emissions generally do not collaborate. According to DOE, DOT, and EPA officials, the three agencies consult on broad issues, such as to discuss available technologies or emissions standards, but these efforts do not involve collaboration on diesel-related issues. Moreover, officials from most of the 14 programs reported that any collaboration across the programs occurs on an informal, case-by-case basis. For example, officials from EPA's Diesel Emissions Reduction Act Program said they may contact officials from the Federal Highway Administration's Congestion Mitigation and Air Quality Improvement program to discuss a specific emissions reduction technology or project that appears in a grant application but that they do not collaborate with officials from this program on a regular basis. Also, some program officials reported that enhanced collaboration could improve the effectiveness of federal funding for activities that reduce diesel emissions. For example, officials from EPA's Diesel Emissions Reduction Act Program said that diesel-related programs could share information to more efficiently award grants and to reduce duplication of agency efforts, such as researching various emissions reduction technologies.

DOE, DOT, and EPA officials generally reported that they do not collaborate on diesel emissions reduction activities with other federal programs because they are unaware of the other programs that fund these activities, including, in some cases, programs within their own agencies. According to agency officials, this is due to the differing purposes and goals of each program, which often do not directly relate to reducing diesel emissions. However, we have previously reported that, although federal programs have been designed for different purposes or targeted for different population groups, coordination among programs with related responsibilities is essential to efficiently and effectively meet national concerns.[22] We reported that uncoordinated program efforts can waste scarce funds, confuse and frustrate program customers, and limit the overall effectiveness of the federal effort. A focus on results as envisioned by the Government Performance and Results Act implies that federal programs contributing to the same or similar results should be closely coordinated to ensure that goals are consistent, and, as appropriate, program efforts are mutually reinforcing. This means that federal agencies are to look beyond their organizational boundaries and coordinate with other agencies to ensure that their efforts are aligned. Also, the Government Performance and Results Act Modernization Act of 2010 requires that agency strategic plans include a description of how the agency is working with other agencies to, among other things, achieve its goals and objectives.

In addition, we have previously reported that agencies face a range of barriers in their efforts to collaborate.[23] To overcome such barriers and to maximize the performance and results of federal programs that share common outcomes, we have previously identified practices that can help agencies enhance and sustain collaboration.[24] These practices include agreeing on agency roles and responsibilities in the collaborative effort and identifying and addressing needs by leveraging collective resources. Further, we have reported that, to the extent that federal efforts are fragmented across agency lines, developing crosscutting performance measures through interagency coordination could ease implementation burdens while strengthening efforts to develop best practices.[25]

CONCLUSION

Over time, EPA has issued more stringent emissions regulations for new diesel engines and vehicles, but existing diesel trucks, buses, locomotives, ships, agriculture equipment, and construction equipment continue to emit harmful pollution. Because diesel engines are durable and energy efficient, it could take decades for these older diesel vehicles and equipment to fall out of use. As a result, federal agencies play an important role in accelerating the attrition of existing diesel engines and vehicles and the resulting reduction in diesel emissions. However, federal funding that reduces diesel emissions is fragmented across 14 programs that overlap in their activities, goals, and eligible recipients. Also, the effectiveness of this funding is unknown because agencies collect limited performance information related to these programs. Because DOT's Federal Transit Administration has not developed performance measures for its goals related to environmental sustainability and economic competitiveness, the agency is unable to fully assess the performance of programs that contribute to these goals. In addition, agencies collect limited information on the results of the diesel emissions reduction activities they fund and do not have a baseline assessment of nationwide diesel emissions, which they could use to measure progress.

Also, collaboration among the 14 programs that fund activities that reduce mobile source diesel emissions is essential to efficiently and effectively reduce diesel emissions. As the focus on results as envisioned by the Government Performance and Results Act implies, federal programs contributing to the same or similar results should be closely coordinated to ensure that goals are consistent, and, as appropriate, program efforts are mutually reinforcing. Agencies often face barriers in their efforts to collaborate, and some best practices for overcoming these barriers include identifying agency roles and responsibilities as well as identifying and leveraging collective resources. Further, when federal efforts are fragmented, this coordination can be achieved through collaboratively developing crosscutting performance measures. However, as we found, these 14 programs generally do not collaborate and collect limited information on the results of the activities they fund that reduce diesel emissions. Without collaboration and crosscutting performance measures, agencies do not have needed information to assess the effectiveness and efficiency of their programs or identify any unnecessary duplication.

Recommendations for Executive Action

We are making two recommendations to help ensure effectiveness and accountability:

1. For transit grant programs, we recommend that the Secretary of Transportation require the Administrator of the Federal Transit Administration to develop quantifiable performance measures, a leading practice based in Government Performance and Results Act principles, for the agency's environmental sustainability and economic competitiveness strategic goals.
2. For federal funding that reduces diesel emissions, we recommend that the Secretary of Energy, the Secretary of Transportation, and the

Administrator of the EPA, consistent with statutory obligations, establish a strategy for collaboration among their grant and loan programs in their activities that reduce mobile source diesel emissions. This strategy should help the agencies

- identify agency roles and responsibilities for activities that reduce diesel emissions, including how a collaborative effort will be led;
- identify and address any unnecessary duplication, as appropriate;
- identify and leverage resources needed to support funding activities that reduce diesel emissions;
- assess baseline levels of diesel pollution and the contributors to mobile source diesel emissions to help agencies target, within their discretion, investments and, as appropriate, inform efforts to measure program effectiveness; and
- develop crosscutting performance measures, as appropriate, to monitor the collective results of federal funding for activities that reduce diesel emissions.

Agency Comments and Our Evaluation

We provided a draft of this report to the Secretary of Energy, the Secretary of Transportation, and the Administrator of EPA for their review and comment. In its written comments, EPA stated that it agreed with our findings and relevant recommendation.

In its comments, DOE questioned several of our findings but agreed with our relevant recommendation. Specifically, DOE stated that our report mischaracterizes the agency as having a statutory responsibility for diesel emissions reductions. Our report does not contain such a statement. Rather, it identifies 14 programs, including 3 DOE programs, that fund activities with the effect of reducing diesel emissions and states that programs with related responsibilities should coordinate their efforts. Our report states that most of the programs we identified have other goals or purposes and do not focus on diesel emissions reduction; nonetheless, each of the programs does fund such activities. Our report also recognizes the varying statutory requirements for each program and recommends that the agencies establish a strategy for collaboration that is consistent with their existing statutory obligations. DOE also stated that our report mischaracterizes DOE as not collaborating with other government agencies. Our report states that DOE collaborates with other agencies on broad issues but does not collaborate on diesel-related issues. In addition, DOE stated that our report mischaracterized the agency as sharing redundant national goals with DOT and EPA. Our report does not discuss DOE's national goals, their relationship to those of other agencies, or whether they are redundant. Rather, our report (1) focuses on DOE programs that fund activities that result in diesel emissions reductions and (2) demonstrates that these programs share similar goals with DOT and EPA programs that fund the same activities. Specifically, each of these programs shares some goals, such as reducing emissions, increasing energy efficiency, and reducing fuel use. DOE also provided technical comments, which we incorporated as appropriate.

DOT questioned several of this report's key findings and its recommendations. Specifically, DOT stated that we inaccurately described the Federal Transit Administration's programs as funding diesel emissions reduction activities. Our report identifies activities that reduce diesel emissions, including replacing existing diesel vehicles and installing devices that reduce idling of diesel engines, and identifies six Federal Transit Administration programs that fund these same activities. In addition, DOT questioned the evidence underlying our finding of fragmentation among the federal programs within our review. DOT stated that we identified independent programs with varying objectives that, in some cases, include similar activities. As we reported, fragmentation occurs when more than one federal agency, or more than one organization within an agency, is involved in the same broad area of national need. Further, our report does not state that fragmentation implies small, incomplete, or broken parts strewn across government, as DOT's comments state. Our report clearly identifies fragmentation, overlap, and duplication among the 14 federal programs that fund diesel emissions reduction activities. Consistent with our established definition of fragmentation and our evidence, we stand by our finding that federal grant and loan funding for activities that reduce diesel emissions is fragmented across 14 programs.

DOT also questioned our finding that the effectiveness of federal funding for diesel emissions reduction activities is unknown. DOT stated that we could have used available air quality data from EPA to assess the effectiveness of the programs we reviewed. We reviewed air quality data from EPA and determined that it was not possible to establish a causal link between the EPA data and the programs we reviewed. Moreover, principles of good governance indicate that agencies that use scarce federal resources should establish quantifiable performance measures for use in administering their programs. This is particularly important when multiple agencies engage in the same or similar activities, even if the activities contribute to different goals. Related to this finding, DOT questioned why the report does not include information that the Federal Transit Administration provided on its contribution to air quality improvement through replacing transit buses. We reviewed this information and found that the numbers the Federal Transit Administration provided were based on unverified assumptions and estimates rather than actual data on the number of diesel buses replaced. As such, the numbers were not reliable for the purposes of our report.

In several instances, DOT questioned our recommendation that the Federal Transit Administration should develop quantifiable performance measures for its environmental sustainability and economic competitiveness strategic goals. DOT's comments on this recommendation reflect a misinterpretation of the recommendation. Specifically, DOT incorrectly stated that our report recommended that the Federal Transit Administration develop quantifiable performance measures relating to diesel emissions reductions. Neither of our recommendations called for DOT to establish such performance measures. Instead, we recommended that (1) the Federal Transit Administration develop performance measures for two of its agencywide strategic goals and (2) DOE, DOT, and EPA establish a strategy for collaboration on diesel emissions reduction activities that, among other things, helps the agencies develop crosscutting performance measures, as appropriate, to assess the collective results of federal funding for activities that reduce diesel emissions. DOT also stated that it operates in full compliance with the Government Performance and Results Act. Specifically, DOT said that it has established outcome-focused performance measures that are appropriate for its programs and mission focus. Our report does not assess DOT's compliance with the Government Performance and Results Act. Rather, it identifies the Federal Transit Administration's strategic goals that relate to the agency's relevant programs and states whether the Federal Transit Administration has developed performance measures for these goals. Our report states that principles of good governance indicate that agencies should establish quantifiable performance measures to demonstrate how they intend to achieve their goals and measure the extent to which they do so. Our report also states that Government Performance and Results Act requirements for agencies to set goals for program performance and to measure results can serve as leading practices for lower levels within federal agencies. We have clarified the report language and recommendation to state that, on the basis of these leading practices, we recommend the Federal Transit Administration establish performance measures for the two agency-wide strategic goals of environmental sustainability and economic competitiveness that relate to the programs involving diesel emissions reduction activities. The Federal Transit Administration provided no evidence that it has established performance measures for these strategic goals. Importantly, the agency's fiscal year 2012 budget justification that it submitted to Congress—the document that Federal Transit Administration officials said contained the agency's goals and performance measures—did not include performance measures for its environmental sustainability and economic competitiveness strategic goals. We continue to believe that the Federal Transit Administration should establish performance measures for these goals.

Regarding our recommendation that DOE, DOT, and EPA establish a strategy for collaboration among their programs that reduce mobile source diesel emissions, DOT agreed that collaboration can be useful but questioned its usefulness in this context. Specifically, DOT stated that the report demonstrates no specific deficiency that has occurred due to the existing level of collaboration. As our report states, DOE, DOT, and EPA were generally unaware of other programs that fund activities that decrease diesel emissions. Additionally, we reported that representatives of several DOE and DOT programs were unaware of related programs within their own agencies that fund the same underlying activities. Our report also states that EPA officials said that enhanced collaboration could improve the effectiveness of federal funding for activities that reduce diesel emissions. In its comments, DOT stated that the report does not offer evidence to support why establishing a strategy for collaboration among entities that fund these activities should be a priority use of federal resources. While the programs we reviewed have been designed for different purposes, coordination among programs with related responsibilities and that fund the same activities is essential to the efficient and effective use of resources. Further, uncoordinated programs can waste scarce funds and limit the overall effectiveness of federal spending. We therefore continue to believe that our recommendation is warranted.

DOT also stated that the report does not effectively demonstrate that our recommended actions will produce cost-effective investments appropriate for DOT that do not potentially duplicate efforts elsewhere in the government. We believe it is entirely appropriate for the Federal Transit Administration to establish performance measures for its goals and do not see how this would duplicate other efforts within the government. We also continue to believe that establishing a strategy for collaboration is an appropriate investment that would help ensure the effectiveness and accountability of federal funding for activities that reduce diesel emissions. As we noted, such a strategy should help agencies identify and address any unnecessary duplication.

Finally, DOT's comments emphasized its view that its programs focus on their statutory mission of transit, whereas diesel emissions reduction is a corollary benefit. Our report states that most of the programs we identified have other goals or purposes and do not focus on diesel emissions reduction; nonetheless, each of the programs does fund such activities. Our report also recognizes the statutory requirements for each program and recommends that the agencies establish a strategy for collaboration that is consistent with existing statutory obligations.

David C. Trimble
Director,
Natural Resources and Environment

APPENDIX I. OBJECTIVES, SCOPE, AND METHODOLOGY

This report examines the (1) extent to which duplication, overlap, fragmentation, or gaps, if any, exist among federal grant, rebate, and loan programs that address mobile source diesel emissions; (2) effectiveness of federal funding for activities that reduce mobile source diesel emissions; and (3) extent to which collaboration takes place among agencies that fund activities that reduce mobile source diesel emissions.

To address the first objective, we identified federal grant, rebate, and loan programs that address mobile source diesel emissions, and reviewed information about each program to identify duplication, overlap, fragmentation, and gaps. To identify the programs that address diesel emissions, we (1) conducted a literature review of government reports, academic materials, legislation, transcripts, appropriations, trade and industry articles, and other relevant publications; (2) interviewed agency officials and relevant industry stakeholders; and (3) reviewed agency documents, including information about activities eligible for funding. For the literature review, we searched twenty databases and websites—Article First, Congressional Research Service, Congressional Budget Office, Inspectors General, Policyfile, ProQuest, Worldcat, National Technical Information Services, Wilson's Applied Science and Technical Abstracts, and the Catalog of Federal Domestic Assistance, among others—for materials published in the last 10 years that may identify relevant federal grant, rebate, and loan programs. Next, we interviewed agency and relevant third-party officials and analyzed agency documents to determine if the programs our searches identified could provide funding for activities that reduce diesel emissions. For this review, we identified programs that fund activities that directly reduce diesel emissions and did not include programs that fund activities, such as research and development efforts, that have the potential to reduce diesel emissions in the future. We held these interviews and conducted these searches from June 2011 to September 2011.

For each program we identified as reducing diesel emissions, we conducted structured interviews of agency officials and reviewed agency documents to determine the types of funding the program provides as well as its purpose, goals, eligible activities, and eligible applicants. We then compared each of these areas across the programs to identify areas of duplication, overlap, or fragmentation. We also compared eligible recipients under each program with available data on the sources of diesel emissions to identify any gaps among the programs, such as mobile sources of diesel emissions for which funding opportunities are not available. For the duplication, overlap, and fragmentation we found, we interviewed agency officials and relevant industry stakeholders to determine its causes and impact.

In addition, we obtained and analyzed funding data from the Department of Energy (DOE), the Department of Transportation (DOT), and the Environmental Protection Agency (EPA) to determine the total amount of federal funding for diesel emissions reduction projects from fiscal years 2007 through 2011. We selected fiscal years 2007 through 2011 as our time period because, by 2007, EPA had issued emissions standards for key on-road sources, such as heavy-duty trucks and buses, as well as a rule requiring refiners to reduce the sulfur content—and therefore the emissions—of certain diesel fuels. We obtained these data from DOE for the Clean Cities, Energy Efficiency and Conservation Block Grant, and State Energy programs; from DOT's Federal Aviation Administration for the Voluntary Airport Low Emissions program; from DOT's Federal Highway Administration for the Ferry Boat and State Infrastructure Banks programs; from DOT's Federal Transit Administration for the National Fuel Cell Bus Technology Development and Transit Investments in Greenhouse Gas Emissions Reduction programs; and from EPA for the Diesel Emissions Reduction Act Program. However, DOT's Federal Transit Administration was unable to provide this data for the Bus and Bus Facilities, Clean Fuels Grant, Congestion Mitigation and Air Quality Improvement, Transit in Parks, and Urbanized Area Formula Grants programs.[26] A Federal Transit Administration official said that because the agency did not track which awards under these programs reduced diesel emissions or diesel fuel use, it was unable to identify the amount of funding each of these programs provided for activities that reduced diesel emissions. The official said that the agency did not track these data because statutory requirements do not call for such tracking. Instead, the agency provided us access to its grants management database, from which we estimated the amount of funding provided for diesel emissions reduction projects for the Clean Fuels Grant, Congestion Mitigation and Air Quality Improvement, and Transit in Parks programs, but we were unable to estimate the amount of funding provided through the Bus and Bus Facilities or Urbanized Area Formula Grants programs due to the limited timeframes of our review. We reviewed documents about the underlying databases that DOE, DOT, and EPA use to collect grant information and interviewed knowledgeable agency officials to assess the reliability of the data for each program. We determined that the data obtained from these agencies were sufficiently reliable for the purposes of this report.

In addition, the Federal Transit Administration provided estimates of the amount that its Bus and Bus Facilities, Clean Fuels Grant, Congestion Mitigation and Air Quality Improvement, Transit in Parks, and Urbanized Area Formula Grants programs awarded from fiscal years 2007 through 2011 for projects that reduced diesel emissions. The agency derived these estimates by identifying obligations made under each of these five programs from fiscal years 2007 through 2011 for purchasing replacement transit vehicles. However, the agency does not consistently collect information on the fuel-type of the vehicles it replaces; rather it collects information on the intended purchase, by fuel-type, for all obligations made in each grant by year and program. The Federal Transit Administration provided this information to GAO; however, this information does not accurately reflect the amount of funding provided for replacement vehicles that reduced diesel emissions, and we did not include the information in this report.

To address the second objective, we reviewed and analyzed agency officials' responses to structured interview questions on their program goals and performance information. We also analyzed agency strategic plans, budget documents, and other agency documentation containing performance information. We reviewed relevant provisions of the Government Performance and Results Act of 1993, as amended by the Government Performance and Results Act Modernization Act of 2010, as well as our prior work on performance measurement.

To address the third objective, we reviewed and analyzed agency officials' responses to structured interview questions on coordination with other programs and assessment of diesel pollution. We also reviewed our prior work on collaboration to compare these programs' efforts with best practices for federal programs. In addition, to identify tax expenditures that provide incentives that address mobile source diesel emissions, we reviewed tax expenditure lists produced by the U.S. Department of the Treasury and the Joint Committee on Taxation; reports by the Congressional Research Service, including the 2010 tax expenditure compendium; and a DOE list of federal incentives related to alternative fuels, vehicles, and air quality.[27] We also interviewed agency officials at DOE, DOT, and EPA as well as industry stakeholders.

We conducted this performance audit from May 2011 to February 2012 in accordance with generally accepted government auditing standards. Those standards require that we plan and perform the audit to obtain sufficient, appropriate evidence to provide a reasonable basis for our findings and conclusions based on our audit objectives. We believe that the evidence obtained provides a reasonable basis for our findings and conclusions based on our audit objectives.

Appendix II. Additional Information on Federal Programs That Provide Grants or Loans for Mobile Source Diesel Emissions Reduction Activities

Department of Energy

Clean Cities program

DOE's Clean Cities program, administered by the Office of Energy Efficiency and Renewable Energy, is a government-industry partnership that works to reduce America's petroleum consumption in the transportation sector. The program provides competitive grants for projects that implement a range of energy-efficient and advanced vehicle technologies, such as hybrids, electric vehicles, plug-in electric hybrids, hydraulic hybrids, and compressed natural gas vehicles, helping reduce petroleum consumption across the United States. The program also supports refueling infrastructure for various alternative fuel vehicles, as well as public education and training initiatives.

Energy Efficiency and Conservation Block Grant program

The Energy Efficiency and Conservation Block Grant program, administered by DOE's Office of Energy Efficiency and Renewable Energy, provides funds through competitive and formula grants to states, territories, federally-recognized Indian tribes, and local governments to develop and implement projects to improve energy efficiency and reduce energy use and fossil fuel emissions in their communities.

State Energy Program

The State Energy Program, administered by DOE's Office of Energy Efficiency and Renewable Energy, provides technical and financial assistance to states through formula and competitive grants. States may use such grants to develop, modify, and implement approved state energy conservation plans.

Department of Transportation

Federal Aviation Administration

Voluntary Airport Low Emissions program

The Voluntary Airport Low Emissions program provides funding to reduce airport ground emissions at commercial service airports in areas failing to meet or maintain National Ambient Air Quality Standards. Grant funding generally supports projects such as electrification of airport gate systems, the incremental cost of purchasing electric luggage carts, and purchasing airport shuttle buses that use alternative fuels. The Federal Aviation Administration considers applications for Voluntary Airport Low Emissions grants on a case-by-case basis based on the project's importance relative to other eligible airport activities. The agency also considers each project's cost effectiveness and reductions in air emissions.

Federal Highway Administration

Congestion Mitigation and Air Quality Improvement program

Jointly administered by Federal Highway Administration and the Federal Transit Administration, the Congestion Mitigation and Air Quality Improvement program provides grants to state departments of transportation, metropolitan planning organizations, and transit agencies for a variety of transportation projects in areas that do not meet or have previously failed to meet federal air quality standards. The program distributes funding through a statutory formula primarily based on population in areas of certain air quality status. The Safe, Accountable, Flexible, and Efficient Transportation Equity Act: A Legacy for Users of 2005 expanded the focus of eligible projects under the program, placing more priority on diesel engine retrofits and cost-effective emission reduction and congestion mitigation projects that also provide air quality benefits.

Ferry Boat Discretionary program

The Intermodal Surface Transportation Efficiency Act of 1991 amended a predecessor ferry program, resulting in this program to construct ferry boats and ferry terminal facilities. Eligible projects include both ferry boats carrying passengers only and those carrying cars and passengers. In general, ferry boats and facilities must be publicly owned or operated, and the ferry facilities must provide connections on a public road, which has not been designated part of the interstate system. The program provides administrative consideration of whether the project will result in a useable facility; what other benefits exist; whether other funds, either state or local, are committed to the project; and whether the project has received program funds in the past.

State Infrastructure Bank program

The State Infrastructure Bank program provides the opportunity to all 50 states, Puerto Rico, the District of Columbia, American Samoa, Guam, the Virgin Islands, and the Commonwealth of the Northern Mariana Islands to establish transportation revolving loan funds. States may capitalize their revolving loan funds with federal highway funding, and states could offer a range of loans and credit options, such as low-interest loans, loan guarantees, or loans requiring repayment of interest-only in early years and delayed repayment of the loan's principal. For example, through a revolving fund, states could lend money to public or private sponsors of transportation projects, project-based or general revenues (such as tolls or dedicated taxes) could be used to repay loans with interest, and the repayments would replenish the fund so that new loans could be supported.

Federal Transit Administration

Bus and Bus Related Equipment and Facilities program

DOT's Bus and Bus Facilities program provides capital assistance for new and replacement buses, related equipment, and related facilities for expansion and maintenance purposes. The projects funded by this program are generally determined by Congress. Funds can be provided only to state and local governmental authorities. The purpose of the program is to replace, rehabilitate, and purchase buses and bus-related facilities in support of FTA's goal of developing a transportation system that (1) maximizes the safe, secure, and efficient mobility of individuals; (2) minimizes environmental impacts; and (3) minimizes transportation-related fuel consumption and reliance on foreign oil.

Clean Fuels Grant program

This program provides competitive grants to assist areas in achieving or maintaining the National Ambient Air Quality Standards for ozone and carbon monoxide and to support emerging clean fuel and advanced propulsion technologies for transit buses and markets for those technologies. Eligible projects under the program include (1) purchasing or leasing clean fuel buses, including buses that employ a lightweight composite primary structure and vans for use in revenue service; (2) constructing or leasing clean fuel bus facilities or electrical recharging facilities and related equipment; and (3) purchasing clean fuel, biodiesel, hybrid electric, or zero emissions technology buses that exhibit equivalent or superior emissions reductions to existing clean fuel or hybrid electric technologies.

National Fuel Cell Bus Technology Development program

This program is a research, development, and demonstration competitive grant program established to facilitate the development of fuel cell bus technology and related infrastructure. The Federal Transit Administration may award grants for this purpose to up to three geographically diverse nonprofit organizations. The goals of the program are to (1) facilitate the development of commercially viable fuel cell bus technologies, (2) significantly improve transit bus fuel efficiency and reduce petroleum consumption, (3) reduce transit bus emissions, (4) establish a globally competitive U.S. industry for fuel cell bus technologies, and (5) increase public acceptance of the fuel cell vehicles.

Paul S. Sarbanes Transit in Parks program

The Transit in Parks program was established to address the challenge of increasing vehicle congestion in and around our national parks and other federal lands by providing competitive grants for capital and planning expenses for new or existing alternative transportation systems in the vicinity of federally owned or managed recreation areas. According to program documents, alternative transportation includes transportation by bus, rail, or any other publicly available means of transportation and includes sightseeing service. It also includes nonmotorized transportation systems such as pedestrian and bicycle trails. The program seeks to conserve natural, historical, and cultural resources; reduce congestion and pollution; improve visitor mobility and accessibility; enhance visitor experience; and ensure access to all, including persons with disabilities.

Transit Investments in Greenhouse Gas and Energy Reduction program

The American Recovery and Reinvestment Act of 2009 authorized the Transit Investments in Greenhouse Gas and Energy Reduction program, and the program received funding through fiscal year 2011. The program did not receive funding for fiscal year 2012 in the relevant appropriations act.[28] This program provides competitive grants to assist public transportation agencies in implementing strategies for reducing greenhouse gas emissions and energy use in transit operations. Eligible applicants under the program include public transportation agencies, federally recognized tribes, and state departments of transportation. Two types of projects are eligible for funding under the Transit Investments in Greenhouse Gas and Energy Reduction program: capital investments that assist in reducing the energy consumption of a transit agency and capital investments that reduce greenhouse gas emissions of a transit agency. For purposes of the Transit Investments in Greenhouse Gas and Energy Reduction program, energy consumption is defined as energy purchased directly by the public transportation agency. Examples of energy include diesel fuel, compressed natural gas, and electricity purchased from power plants. Emissions are defined as those emitted directly by the assets of the public transportation agency.

Urbanized Area Formula Grants program

This program provides grants to urbanized areas and to states for public transportation capital projects and operating assistance for equipment and facilities in urbanized areas and for transportation-related planning.[29] The program allocates funds based on a multitiered formula, which separates urban areas with populations under 200,000 from those with populations of 200,000 or more. Funds are first apportioned based on a formula provided in law to designated recipients (typically metropolitan planning organizations or a state or regional authority responsible for capital projects and for financing and directly providing public transportation). Designated recipients allocate the apportionment among eligible transit service providers in the urbanized area. Eligible uses of program funds include planning, design, and evaluation of transit projects and capital investments in bus-related activities, such as replacement, overhaul, and rebuilding of buses.

Environmental Protection Agency

Diesel Emissions Reduction Act Program

This program provides grant funding to reduce emissions from existing diesel engines through engine retrofits, rebuilds, and replacements; switching to cleaner fuels; and other strategies. The program offers funding through four subprograms:

- the National Clean Diesel Funding Assistance Program awards competitive grants for projects implementing EPA verified and certified diesel emissions reduction technologies,
- the National Clean Diesel Emerging Technologies Program awards competitive grants for projects that develop and evaluate emerging diesel emissions reduction technologies,
- the SmartWay Clean Diesel Finance Program awards competitive grants to establish low-cost revolving loans or other innovative financing programs that help fleets reduce diesel emissions, and
- the State Clean Diesel Grant Program allocates funds to participating states to implement grant and loan programs for clean diesel projects.

APPENDIX III. TAX EXPENDITURES RELATED TO MOBILE SOURCE DIESEL EMISSIONS REDUCTION ACTIVITIES

Excise Tax Exemption for Idling Reduction Devices

This tax expenditure excludes certain idling reduction devices from the federal excise tax. Under federal excise tax law, heavy truck, trailer, and tractor parts sold separately from the vehicle generally are subject to a 12 percent retail tax. The Energy Improvement and Extension Act of 2008 excludes qualified idling reduction devices from the federal retail tax on vehicle parts. EPA, in consultation with the Secretaries for the DOT and DOE, maintains a list of devices approved for the tax exemption. An idle reduction device is generally a device or system that provides services, such as heat, air conditioning, or electricity, to the vehicle or equipment without the use of the main drive engine while the vehicle or equipment is temporarily parked or remains stationary, hence reducing unnecessary idling of the vehicle or equipment. No estimate of forgone federal tax revenue for this excise tax provision is available because the Department of the Treasury reports estimates only for income tax expenditures and does not report estimates for tax provisions that result in forgone excise tax only.

Biodiesel and Small Agri-biodiesel Producer Tax Credits

This tax expenditure provides an income tax credit as well as an excise tax credit for the production and use of biodiesel.[30] The use of biodiesel instead of conventional diesel fuel significantly reduces particulate matter and hydrocarbon emissions. The biodiesel fuels income tax credit is the sum of three credits: (1) the biodiesel mixture credit, which provides $1 for each gallon of biodiesel and agri-biodiesel used by the taxpayer in the production of a qualified biodiesel mixture; (2) the biodiesel credit, which is $1 per gallon for each gallon of unblended biodiesel and agri-biodiesel when used as a fuel or sold at retail; and (3) the small agri-biodiesel producer credit, which is 10 cents per gallon for up to 15 million gallons of agri-biodiesel produced by small producers. The biodiesel excise tax credit provides a tax credit of $1 for each gallon of biodiesel or agri-biodiesel a taxpayer used to produce a biodiesel mixture for sale or use in a trade or business. Renewable diesel fuel is eligible for both the income tax credit and excise tax credit at a rate of $1 per gallon. According to Department of the Treasury estimates, in fiscal year 2010, the biodiesel income tax credits resulted in $20 million in forgone federal income tax revenue, and the biodiesel excise tax credit resulted in $490 million in forgone federal excise tax revenue.

Diesel Fuel Emulsion Tax Credit

A claim for credit or refund may be made for the nontaxable use of a diesel-water fuel emulsion—a mixture of diesel, water, and additives—and for undyed diesel fuel used to produce a diesel-water fuel emulsion. The presence of water in the emulsion reduces both nitrogen oxide and particulate matter emissions from the diesel fuel. The claim rate for nontaxable use of a diesel-water fuel emulsion taxed at 19.8 cents per gallon is 19.7 cents (if exported, the claim rate is 19.8 cents). The following are the nontaxable uses for a diesel-water fuel emulsion for which a credit or refund may be allowable to an ultimate purchaser: on a farm for farming purposes; off-highway business use; export; in a qualified local bus; in a school bus; other than as fuel in the propulsion engine of a train or diesel-powered highway vehicle, but not off-highway use; exclusive use by a qualified blood collector organization; in a highway vehicle owned by the United States that is not used on a highway; exclusive use by a nonprofit educational organization; exclusive use by a state, political subdivision of a state, or the District of Columbia; and in an aircraft or vehicle owned by an aircraft museum. No estimate of forgone federal tax revenue for this excise tax provision is available because the Department of the Treasury reports estimates only for income tax expenditures and does not report estimates for tax provisions that result in forgone excise tax only.

APPENDIX IV. STRATEGIC GOALS AND ESTABLISHMENT OF PERFORMANCE MEASURES RELATED TO PROGRAMS THAT FUND ACTIVITIES THAT REDUCE MOBILE SOURCE DIESEL EMISSIONS

Agency	Strategic goal	Performance measure established	Related programs
DOE	To catalyze the timely, material, and efficient transformation of the nation's energy system and secure U.S. leadership in clean energy technologies	Yes	Clean Cities program
			Energy Efficiency and Conservation Block Grant program
			State Energy Program
DOT			
Federal Aviation Administration	To advance environmentally sustainable policies and investments that reduce carbon and other harmful emissions from transportation sources	Yes	Voluntary Airport Low Emissions program
	To ensure the United States proactively maintains its critical transportation infrastructure in a state of good repair	Yes	

(Continued)

Agency	Strategic goal	Performance measure established	Related programs
	To improve public health and safety by reducing transportation-related fatalities and injuries	Yes	
	To promote transportation policies and investments that bring lasting and equitable economic benefits to the nation and its citizens	Yes	
Federal Highway Administration	To advance environmentally sustainable policies and investments that reduce carbon and other harmful emissions from transportation sources	Yes	Congestion Mitigation and Air Quality Improvement program
			Ferry Boat Discretionary program
	To ensure the United States proactively maintains its critical transportation infrastructure in a state of good repair	Yes	Ferry Boat Discretionary program
	To foster livable communities through place-based policies and programs that increase transportation choices and access to transportation services	Yes	Congestion Mitigation and Air Quality Improvement program
			Ferry Boat Discretionary program
	To improve public health and safety by reducing transportation-related fatalities and injuries	Yes	Congestion Mitigation and Air Quality Improvement program
			Ferry Boat Discretionary program
	To promote transportation policies and investments that bring lasting and equitable economic benefits to the nation and its citizens	Yes	Ferry Boat Discretionary program
Federal Transit Administration	To advance environmentally sustainable policies and investments that reduce carbon and other harmful emissions from transportation sources	No	Transit in Parks program

Agency	Strategic goal	Performance measure established	Related programs
	To foster livable communities through place-based policies and programs that increase transportation choices and access to transportation services	Yes	Bus and Bus Facilities program
			Clean Fuels Grant program
			Transit in Parks program
			Urbanized Area Formula Grants program
	To improve public health and safety by reducing transportation-related fatalities and injuries	Yes	Bus and Bus Facilities program
			Clean Fuels Grant program
			Transit in Parks program
			Urbanized Area Formula Grants program
	To promote transportation policies and investments that bring lasting and equitable economic benefits to the nation and its citizens	No	Bus and Bus Facilities program
			Clean Fuels Grant program
			Transit in Parks program
			Urbanized Area Formula Grants program
EPA	To reduce greenhouse gas emissions and develop adaptation strategies to address climate change, and protect and improve air quality	Yes	Diesel Emissions Reduction Act Program

Source: GAO analysis of DOE, DOT, and EPA documents.

End Notes

[1] Nitrogen oxides are regulated pollutants commonly known as NOx that, among other things, contribute to the formation of ozone. Particulate matter is an ubiquitous form of air pollution commonly referred to as soot.

[2] Nonroad engines are those used in machines, such as construction equipment, agricultural equipment, and airport service vehicles. Also, EPA does not maintain information on the number of mobile sources of diesel emissions built after 2007.

[3] GAO, *Managing for Results: Using the Results Act to Address Mission Fragmentation and Program Overlap*, GAO/AIMD-97-146 (Washington, D.C.: Aug. 29, 1997). For more

information on fragmentation, overlap, and duplication in federal programs see GAO, *Opportunities to Reduce Potential Duplication in Government Programs, Save Tax Dollars, and Enhance Revenue*, GAO-11-318SP (Washington, D.C.: Mar. 1, 2011).

[4] GAO/AIMD-97-146.

[5] Pub. L. No. 111-364 § 3(a)-(b). 124 Stat. 4060 (2011). Additionally, in this report, we use the term "diesel emissions" when referring to both diesel emissions and diesel fuel use.

[6] Tax expenditures include exemptions and exclusions from taxation, deductions, credits, deferral of tax liability, and preferential tax rates. The revenue that the government forgoes in these instances may be viewed as spending channeled through the tax system.

[7] The Government Performance and Results Act Modernization Act of 2010, Pub. L. No. 111-352, 124 Stat. 3866 (2011), amends the Government Performance and Results Act of 1993, Pub. L. No. 103-62, 107 Stat. 285 (1993).

[8] These standards are expressed as concentration limits in the ambient air averaged over a specific period—such as 1 hour or 8 hours—and compliance is determined through localized ground level monitoring.

[9] Nondiesel mobile sources, industrial processes, and power plants, among other sources, generated the remaining nitrogen oxide emissions; fossil fuel combustion, dust, and agricultural activities, among other sources, generated the remaining particulate matter emissions.

[10] EPA also sets standards for new or upgraded stationary sources of diesel emissions, such as engines used to generate electricity at power and manufacturing plants. In 2008, EPA estimated that more than 900,000 stationary diesel engines were in operation.

[11] Competitive grant programs award funds for specific projects or activities based on eligibility and selection criteria as established by law or regulation, or on an administrative basis. Formula grant programs allocate funds to states or their subdivisions in accordance with a distribution formula prescribed in law or administrative regulation. Grant recipients may then allocate these funds to specific projects based on program eligibility guidelines.

[12] Under DOT's State Infrastructure Banks program, states may use allocated federal transportation funds to capitalize state infrastructure banks, which in turn provide loans and other nongrant financial assistance to eligible projects.

[13] All dollar amounts in this report are in nominal dollars.

[14] DOT's Federal Transit Administration was unable to provide this information for the Bus and Bus Facilities program; Clean Fuels Grant program; Congestion Mitigation and Air Quality Improvement program, for which the Federal Transit Administration manages grant data; the Transit in Parks program; and the Urbanized Area Formula Grants program. According to agency officials, the agency does not track which grants fund projects that reduce diesel emissions, and agency officials said that they would have to perform a labor-intensive, time-consuming review—diverting limited resources at the critical end of the fiscal year period—to estimate the amount of program funding that has reduced diesel emissions. Instead, the agency provided us with access to its grants database, from which we estimated the amount of funding provided for diesel emissions reduction projects for the Clean Fuels Grant, Congestion Mitigation and Air Quality Improvement, and Transit in Parks programs, but we were unable to estimate the amount of funding provided through the Bus and Bus Facilities or Urbanized Area Formula Grants programs due to the time frame of our review. See appendix I for additional information.

[15] GAO/AIMD-97-146 and GAO, *Agencies' Annual Performance Plans Under the Results Act: An Assessment Guide to Facilitate Congressional Decisionmaking*, GAO/GGD/AIMD-10.1.18 (Washington, D.C.: February 1998).

[16] GAO/AIMD-97-146.

[17] See GAO, *Environmental Justice: EPA Needs to Take Additional Actions to Help Ensure Effective Implementation*, GAO-12-77 (Washington, D.C.: Oct. 6, 2011).

[18] See GAO/GGD/AIMD-10.1.18 and GAO, *The Results Act: An Evaluator's Guide to Assessing Agency Annual Performance Plans*, GAO/GGD-10.1.20 (Washington, D.C.: April 1998).

[19] Americans with Disabilities Act of 1990. Pub. L. No. 101-336, 104 Stat. 327, codified as amended at 42 U.S.C. §§ 12101 et seq.

[20] EPA calculated these emissions reduction figures in tons of emissions reduced over the lifetime of each project funded. Also, EPA collects information on emissions reductions but has not reported data more current than fiscal year 2008.

[21] We have previously reported that, in general, agencies should focus their funding on areas of greatest need. See GAO, *21st Century Challenge: Reexamining the Base of the Federal Government*, GAO-05-325SP (Washington, D.C.: February 2005).

[22] GAO, *The Government Performance and Results Act: 1997 Governmentwide Implementation Will Be Uneven*, GAO/GGD-97-109 (Washington, D.C.: June 2, 1997).

[23] GAO, *Managing For Results: Barriers to Interagency Coordination*, GAO/GGD-00-106 (Washington, D.C.: Mar. 29, 2000).

[24] GAO, *Results-Oriented Government: Practices That Can Help Enhance and Sustain Collaboration among Federal Agencies*, GAO-06-15 (Washington, D.C.: Oct. 21, 2005).

[25] GAO/AIMD-97-146.

[26] DOT's Federal Highway Administration administers the Congestion Mitigation and Air Quality Improvement program, but DOT's Federal Transit Administration collects and maintains grant data for the program.

[27] The published lists report federal income tax expenditures and also include information on excise tax revenue forgone for listed tax expenditures. The lists do not include estimates for tax provisions that result in forgone excise tax only.

[28] See H.R. Rep. No. 112-284, at 305 (Conf. Rep.).

[29] An urbanized area is an area with a population of 50,000 or more that is designated as such in the 2000 Census by the U.S. Department of Commerce, Bureau of the Census.

[30] Biodiesel fuel is an alternative to petroleum-based transportation fuel. U.S. biodiesel is made from soybeans and other plant oils, such as cottonseed and canola; animal fats, such as beef tallow, pork lard, and poultry fat; and recycled cooking oils.

In: Diesel Emissions: Reduction Efforts ISBN: 978-1-62081-561-8
Editors: J. F. Price and S. E. Mitchell

Chapter 3

THE COST-EFFECTIVENESS OF HEAVY-DUTY DIESEL RETROFITS AND OTHER MOBILE SOURCE EMISSION REDUCTION PROJECTS AND PROGRAMS*

United States Environmental Protection Agency

INTRODUCTION

The Safe, Accountable, Flexible, Efficient Transportation Equity Act: A Legacy for Users (SAFETEA-LU) was enacted August 10, 2005, as Public Law 109-59. SAFETEA-LU creates new opportunities as well as responsibilities for States, Metropolitan Planning Organizations (MPOs), and other agencies involved in the selection of projects funded through the Congestion Mitigation and Air Quality Improvement Program (CMAQ). Among the changes, Section 1808 of SAFETEA-LU, for the first time, makes CMAQ funding available for many nonroad diesel retrofits for nonroad vehicles, engines, and construction equipment that are used on Title 23 projects in certain nonattainment areas. While acknowledging State and MPO roles and authority for making final project selections, Section 1808 also addresses funding priority. "(A) IN GENERAL – States and metropolitan planning

* This is an edited, reformatted and augmented version of the Transportation and Regional Programs Division, U.S. Environmental Protection Agency, EPA420-B-07-006, dated May 2007.

organizations shall give priority in distributing funds received for congestion mitigation and air quality projects and programs from apportionments derived from applications of sections 104(b)(2)(B) and 104(b)(2)(C) to – (i) diesel retrofits, particularly where necessary to facilitate contract compliance, and other cost-effective emission reduction activities, taking into consideration air quality and health effects; and (ii) cost-effective congestion mitigation activities that provide air quality benefits. (B) SAVINGS – This paragraph is not intended to disturb the existing authorities and roles of governmental agencies in making final project selections." To support the implementation of Section 1808, Congress included a provision for the EPA, in consultation with DOT, to publish information on diesel retrofit technologies that have been: 1) certified or verified by EPA or the California Air Resources Board (CARB)[1]; 2) identified by EPA or CARB as having an application and approvable test plan for verification; and 3) other available information regarding the emission reduction effectiveness and cost-effectiveness of diesel emission control technologies, taking into consideration air quality and health effects.[2] In addition, The Energy Policy Act of 2005 (EPACT), enacted as Public Law 109-58 includes a Diesel Emissions Reduction Program that authorizes funding to establish cost-effective clean diesel projects.[3] This document will assist eligible entities with designing and implementing projects and will fulfill EPA's obligation under EPACT to provide information regarding the cost-effectiveness of eligible technologies for reducing emissions.[4] EPA is issuing this document in fulfillment of these statutory requirements. States and local agencies involved with developing and implementing air pollution control programs should consult this information when making choices about investments to achieve air quality and public health goals. This document does not substitute for any applicable regulations, nor is it a regulation itself. It does not impose binding, enforceable requirements on any party and may not apply to a particular situation based upon the circumstances. EPA and State decision makers, as well as other interested parties are free to raise questions about the applicability of this information to a particular situation. The information in this document may be revised periodically without public notice as new data and research becomes available.

BACKGROUND

In previous transportation legislation, Congress required the National Academy of Sciences (NAS) to evaluate the CMAQ Program. The results of the NAS CMAQ study demonstrate that certain CMAQ eligible activities can be far more cost-effective in generating emission reductions than others.[5] Furthermore, implementation of the 8-hour ozone and fine particulate matter ($PM_{2.5}$) National Ambient Air Quality Standards (NAAQS) underscores a need, in some areas, to identify additional local emission reduction activities that can help nonattainment areas meet the earliest of the NAAQS attainment dates (2010).

Retrofit projects can begin producing emission reductions immediately and can help State and local governments reduce emissions of $PM_{2.5}$, nitrogen oxides (NOx), and volatile organic compounds (VOC) in the near term. Retrofits include a wide range of emission reduction strategies available for diesel vehicles and equipment, including:

- Retrofitting engines with verified technologies
- Using cleaner fuels
- Replacing older equipment
- Repowering (replacing old engines with new, cleaner engines
- Reducing idling
- Properly maintaining equipment
- Gaining operational efficiencies

Retrofit technologies are advancing at a rapid pace. The use of established technologies, such as diesel oxidation catalysts (DOCs) and diesel particulate filters (DPFs), continues to grow exponentially, while new, emerging technologies such as Lean NOx (LNC) catalysts, are steadily improving. Retrofit technologies often vary in the type of pollutant reduced. DOCs and DPFs remove PM from the exhaust, but do not reduce NOx. However, DOCs or DPFs can be combined with a NOx reduction strategy – such as a cleaner fuel – to enhance the emission reduction benefits.

While retrofit technologies are one option for reducing diesel emissions, other options include cleaner fuels such as compressed natural gas (CNG) and the replacement of older engines and equipment. Cleaner fuels are becoming more prevalent throughout the country. The switch to ultra-low sulfur diesel (ULSD) fuel for highway engines enables advanced emissions reduction technologies to operate effectively (e.g. DPFs). Another option that can be applied to any vehicle or equipment is to reduce idling. Simply turning off the engine when the vehicle or machine is not in use can reduce emissions as well as save fuel and minimize wear and tear on the engine.

As described below, diesel retrofits are especially important in areas facing immediate challenges to reduce $PM_{2.5}$ emissions and minimize the health impacts of diesel exhaust on its citizens. Some retrofits can provide benefits in the near-term by accelerating the use of new emission reduction technologies on existing older vehicles and engines. Otherwise, these vehicles and engines would continue emitting at higher rates throughout their useful lives. Given their immediate benefits, State and local agencies should work together to determine how retrofit projects could best be used, given local air quality characteristics. One option is to re-evaluate the programming of CMAQ funds with an emphasis on supporting $PM_{2.5}$ emission reduction strategies that can cost-effectively help areas attain the $PM_{2.5}$ NAAQS in 2010, reduce $PM_{2.5}$ health risks and meet their transportation conformity requirements.[6]

While SAFETEA-LU emphasizes eligibility and priority funding for retrofits, it is important to note that the CMAQ program is not the only funding resource available to support retrofit programs. Collaborations of federal, state, and local government agencies, nonprofit organizations, and industry, are working together to fund retrofit projects across the country. Information about these collaboratives and funding opportunities are available at: http://www.epa.gov/cleandiesel/grantfund.htm

HEALTH IMPACTS

Reducing emissions from diesel engines is one of the most important public health challenges facing the country. EPA recently finalized two sets of clean fuel and vehicle emissions standards that will lead to dramatic emission reductions in new diesel-powered engines. Included within these rulemakings are cleaner fuel requirements, such as the use of ultra-low sulfur diesel, which will provide immediate emissions reductions in both new and older diesel engines.

Even with more stringent heavy-duty highway and nonroad engine standards set to take effect over the next decade, millions of diesel engines already in use will continue to emit large amounts of $PM_{2.5}$ emissions, which contribute to serious public health problems. These emissions cause thousands of premature deaths, hundreds of thousands of asthma attacks, millions of lost work days, and numerous other health impacts every year. There are almost 65 million people living in 120 counties with monitored $PM_{2.5}$ levels (2000–2002) exceeding the $PM_{2.5}$ NAAQS.

Diesel engines emit large amounts of nitrogen oxides and particulate matter. In 2004, diesel engines produced more than 6.2 million tons of NO_x and more than 300,000 tons of $PM_{2.5}$. Highway diesel engines accounted for approximately 50% of NO_x and 30% of $PM_{2.5}$ from the mobile source diesel sector. The freight sector's contribution to diesel emissions – which included rail movement for the first time in 2004 – was approximately 56% of NOx and 32% of $PM_{2.5}$. Nonroad equipment accounted for 24% of NO_x and almost half of diesel $PM_{2.5}$ emissions in 2004. While marine engines account for only 3% of diesel engines in 2004, port and non-port related marine emissions accounted for 16% of NO_x and 14% of PM of all mobile diesel sources.

EPA believes that diesel exhaust is likely to be carcinogenic to humans. The risk associated with exposure to diesel exhaust includes the particulate and gaseous components among which are benzene, formaldehyde, acetaldehyde, acrolein, and 1, 3–butadiene, all of which are known or suspected human or animal carcinogens, or have noncancer health effects.[7] Specifically diesel exhaust has been judged to pose a lung cancer hazard for humans as well as a hazard from noncancer respiratory effects such as pulmonary inflammation. EPA assessed air toxic emissions and their associated risk (the National-Scale Air Toxics Assessment or NATA for 1999), and we concluded that diesel exhaust ranks with other substances that the national-scale assessment suggest pose the greatest relative risk.

Health effects associated with short-term variation in ambient PM have been indicated by epidemiologic studies showing associations between exposure and increased hospital admissions for ischemic heart disease, heart failure, respiratory disease, including chronic obstructive pulmonary disease (COPD) and pneumonia. Short-term elevations in ambient PM have also been associated with increased cough, lower respiratory symptoms, and decreases in lung function. Additional studies have associated changes in heart rate and/or heart rhythm in addition to changes in blood characteristics with exposure to ambient PM. Short-term variations in ambient PM have also been associated with increases in total and cardiorespiratory mortality.[8] Children are especially sensitive to diesel emissions compared to healthy adults because their respiratory systems are still developing and they have a faster breathing rate. Areas that are attaining the $PM_{2.5}$ standard may wish to consider these factors when planning on how to effectively remain in attainment and protect the public health.

Diesel exhaust also contributes to the formation of ground level ozone. While much has been accomplished in reducing ozone levels, ground-level ozone remains a pervasive pollution problem in many areas of the United

States. Exposure to ozone has been linked to a number of health effects, including significant decreases in lung function, inflammation of the airways, and increased respiratory symptoms, such as cough and pain when taking a deep breath. Exposure can also aggravate lung diseases such as asthma, leading to increased medication use and increased hospital admissions and emergency room visits. Active children are the group at highest risk from ozone exposure because they often spend a large part of the summer playing outdoors. Children are also more likely to have asthma, which may be aggravated by ozone exposure. Other at-risk groups include adults who are active outdoors (e.g., some outdoor workers) and individuals with lung diseases such as asthma and chronic obstructive pulmonary disease. In addition, long-term exposure to moderate levels of ozone may cause permanent changes in lung structure, leading to premature aging of the lungs and worsening of chronic lung disease.[9]

Ozone also affects vegetation and ecosystems, leading to reductions in agricultural crop and commercial forest yields, reduced growth and survivability of tree seedlings, and increased plant susceptibility to disease, pests, and other environmental stresses (e.g., harsh weather). In long-lived species, these effects may become evident only after several years or even decades and may result in long-term effects on forest ecosystems. Ground level ozone injury to trees and plants can lead to a decrease in the natural beauty of our national parks and recreation areas.

COST-EFFECTIVE EMISSION REDUCTION ACTIVITIES

Cost-effectiveness, for the purpose of this document, is defined as the cost per ton of emissions reduced. Within this context, cost-effectiveness can vary depending on a number of factors.

The pollutant(s) for which the area is in nonattainment, precursor pollutants of concern, relative size of pollutant inventories, and the existing sources and level of control measures in place can all influence cost-effectiveness. It is also important to note that cost-effectiveness does not necessarily correspond with overall effectiveness.

For instance, a certain project or technology may be very cost-effective at reducing VOCs in an ozone nonattainment area, but if the project or technology only applies to very few emissions sources, or if the air quality chemistry in the ozone nonattainment area is NO_x dependant, the overall effectiveness in reducing ozone may be quite limited.

Where essential emission reductions are more difficult to achieve, the acceptable cost of achieving those reductions could increase. Areas with more serious air quality problems typically will need to obtain greater levels of emissions reductions from local sources than areas with less serious problems, and it would be expected that their residents could realize greater health benefits from such reductions.

It may be reasonable and appropriate for areas with more serious air quality problems to fund emission reduction requirements with generally higher costs per ton than the cost of emissions reductions in areas with less serious air quality problems.

Given these considerations, EPA believes that it is not necessary to propose a fixed dollar per ton cost threshold for identification of cost-effective emission reduction activities.

Retrofitting diesel engines is, however, one of the most cost-effective ways to reduce diesel emissions. The term diesel retrofit includes any technology or system that achieves emission reductions beyond that required by the EPA regulations at the time of new engine certification.

Diesel retrofit projects include the replacement of high-emitting vehicles/equipment with cleaner vehicles/equipment (including hybrid or alternative fuel models), repowering or engine replacement, rebuilding the engine to a cleaner standard, the purchase and installation of advanced emissions control technologies (such as particulate matter traps or oxidation catalysts) or the use of a cleaner fuel. For example, diesel oxidation catalysts can result in a per vehicle particulate matter reductions of 20-40%. Diesel particulate filters reduce particulate matter up to 90% per vehicle.

To help stakeholders identify cost-effective technologies, EPA has developed a list of verified retrofit technologies that contains information on expected emission reduction benefits.

This list provides information on numerous innovative emission control technologies that EPA has approved for receiving emission reduction credit.

Each EPA verified technology has undergone extensive testing and analysis. The verification process includes evaluations of the emissions reduction performance of retrofit technologies- including the durability of the technologies- and identification of engine operating criteria and other conditions that must exist for these approved technologies to achieve the verified level of reductions.

EPA evaluates each technology using a specific fuel, on a specific engine, and under specific loading cycles. The California Air Resources Board (CARB) has a verification process similar to EPA's verification process. EPA has signed a Memorandum of Agreement with CARB to recognize CARB's list of verified emission control options.[10]

In addition, EPA has established a comprehensive list of idle-control technologies, which is available on the EPA website.[11]

To help stakeholders compare cost-effective strategies, EPA has included an appendix with 4 tables containing estimates of the cost per ton of pollutant reduced, for projects and programs that are potentially eligible for CMAQ funding. The estimates are derived from the best data available to EPA at the time this document was issued. The source of the information is identified for each project category.

The tables include cost-effectiveness estimates for reducing NO_x and VOC precursor emissions for ozone and for PM emissions. As noted above, it is not always constructive to do a direct comparison between the cost-effectiveness of reducing different pollutants.

For instance, PM and NO_x cost-effectiveness are not comparable because the health effects, emissions inventories and control sources for the two pollutants are very different. Generally, emissions inventories show much greater amounts of NO_x compared to PM. Correspondingly, greater reductions of NO_x emissions are required to reduce ambient ozone levels than reductions of PM emissions required to reduce ambient PM levels.

While reducing a ton of PM often costs more than to reduce a ton of NO_x, the health effects of PM are greater per ton than for NO_x. In addition to assessing the cost-effectiveness of reducing a pollutant, careful consideration should be given to the overall effectiveness of the reductions. Due to the greater health hazard posed by PM, a little PM reduction may be more effective than larger NO_x reductions from a public health perspective.

Table 1 summarizes PM cost-effectiveness for typical diesel retrofit scenarios that utilize a diesel oxidation catalyst (DOC) or catalyzed diesel particulate filter (CDPF). These cost-effectiveness estimates have not factored in the co-benefits from reducing other pollutants such as VOC.

The cost-effectiveness of retrofitted programs can vary significantly depending on a number of factors, including actual annual average activity (i.e., annual vehicle miles traveled for highway or annual operating hours for nonroad).

Table 2 is similar to Table 1 and contains PM cost-effectiveness for typical nonroad retrofit projects utilizing a diesel oxidation catalyst (DOC) or catalyzed diesel particulate filter (CDPF).

More information about the data and the methodology used to develop these tables is in EPA's technical report entitled "Diesel Retrofit Technology:

An Analysis of the Cost Effectiveness of Reducing Particulate Matter Emissions from Heavy-Duty Diesel Engines Through Retrofits." This report can be found at: http://www.epa.gov/ cleandiesel/publications.htm

Table 3 summarizes the cost-effectiveness of some specific voluntary mobile source emission reduction programs supported by EPA's Office of Transportation and Air Quality.

These voluntary programs aim to achieve cost-effective emission reductions without the need for regulation. These programs are implemented through partnerships with small and large businesses, citizen groups, industry, manufacturers, trade associations, and state and local governments. More information about these programs can be found at: http://epa.gov/otaq /voluntary.htm

Table 4 contains project and program categories taken from the study by the NAS in 2002. The NAS CMAQ study assessed the cost-effectiveness of various CMAQ-eligible strategies to reduce emissions. The study estimated the cost-effectiveness of projects based on cost (in calendar year 2000 dollars) per ton of emissions (VOC and NOx) reduced.

To bring the cost estimates current, EPA adjusted the NAS's estimates to 2006 dollars according to the consumer price index established by the Bureau of Labor Statistics, U.S. Department of Labor.

Recognizing that ozone levels in many nonattainment areas are more dependant on NOx than on VOCs, and consistent with the NAS methodology, the Table 4 estimates assume a benefit weighting ratio of 4:1 for NOx:VOC. This approach to weighting is commonly used to calculate one cost-effectiveness value for multiple pollutant reductions.

The weighting factors are determined on the basis of relative damage values of the individual pollutants. The damage value of a given pollutant is estimated through modeling of air quality, assessing human and environmental exposure, and then applying a valuation to the health and environmental effects of the exposure.

It is important to note that while the estimates reflect the best available data at the time, there are limitations inherent in such an assessment. The data presented are based on a select sampling of projects that may not completely capture the potential cost-effectiveness of other techniques of implementing particular strategies.

Therefore, the median cost should be considered along with the cost range to better portray a project's potential cost-effectiveness. Further analysis is recommended in order to assess the cost-effectiveness and emission reduction potential of specific projects or programs. EPA provides models for estimating the transportation sector's impacts on emissions. Emissions estimates from these models can be used for evaluating the impact and cost-effectiveness of various emissions control strategies.[12]

NATIONAL AMBIENT AIR QUALITY STANDARDS

Consistent with the statutory provisions of SAFETEA-LU regarding CMAQ funding priority for diesel retrofits, implementation of the NAAQS for $PM_{2.5}$ and 8-hour ozone also highlights the critical need to target CMAQ funds to diesel engine retrofits and the most cost-effective emission reduction activities. While substantial emission reductions have been achieved in the mobile source sector, recent data show that mobile sources remain a major source of air pollution in most nonattainment areas.[13]

Many States and MPOs with $PM_{2.5}$ nonattainment areas face a unique challenge in that their CMAQ funds have been previously allocated to support attainment of the ozone NAAQS.

The initial statutory attainment date for the current $PM_{2.5}$ NAAQS is as expeditiously as practical but no later than 2010 and emission reductions needed to attain the standard must be in place at the beginning of 2009.[14] States are required to submit $PM_{2.5}$ state implementation plans (SIPs) in 2008 to demonstrate how the nonattainment areas will attain the $PM_{2.5}$ NAAQS. Therefore, it is particularly urgent that States and MPOs in $PM_{2.5}$ areas begin now to direct CMAQ resources to the control of diesel emissions.

States and MPOs in 8-hour ozone nonattainment areas should also consider how CMAQ resources can be used to fund retrofit projects and other cost-effective measures to help attain the 8-hour ozone NAAQS. Eight-hour ozone nonattainment areas have varying attainment dates depending on the level of their ozone concentrations.

Some areas may need to implement local control measures to achieve emission reduction in advance of the timeframe for emission reductions projected to be achieved by national control measures. States and MPOs are also required to meet the transportation conformity requirements of section 176(c) of the CAA.

Transportation conformity ensures that emissions that result from an area's transportation system stay within the limits established in the SIP. In some $PM_{2.5}$ and 8-hour ozone areas, staying within these emissions limits may necessitate additional controls, potentially including diesel emission reduction strategies.

On September 21, 2006, EPA issued new standards which strengthen the 24-hour $PM_{2.5}$ standard. EPA anticipates that attainment (meeting the standards) and nonattainment (violating the standards) designations for areas with respect to the new standard will become effective in April 2010.[15] States with nonattainment areas must submit plans by April 2013 to demonstrate how they will attain the standards by 2015.

While the most pressing need for the states is to develop local measures to meet the current $PM_{2.5}$ standard, the further tightening of the $PM_{2.5}$ standard and the retention of the 24-hour coarse PM standard may create an additional need for reductions in the period from 2010 to 2015.

For these reasons, EPA strongly recommends that States and MPOs re-evaluate their current and proposed transportation plans and CMAQ activities to ensure an appropriate balance in projects to support attainment of the $PM_{2.5}$ and 8-hour ozone NAAQS.

EPA Emission Reduction Guidance and Cost Effectiveness Resources

EPA, in consultation with DOT and stakeholders, has developed several guidance documents to help MPOs and others take emission reduction credit for CMAQ (or other) funded activities that retrofit diesel engine trucks, nonroad equipment (such as construction and locomotives), school buses, reduce idling from diesel trucks, and support strategies to reduce drive-alone commutes.

In addition, EPA has released quantification tools for estimating the environmental impacts and cost effectiveness of emission reduction technologies to vehicles and equipment. These documents and resources are:

- U.S. EPA. Diesel Retrofits: *Quantifying and Using Their Benefits in SIPs and Conformity. Guidance for State and Local Air and Transportation Agencies*. EPA420-B-06-005. http://www.epa.gov/otaq/stateresources/transconf/policy/420b06005.pdf
- U.S. EPA. *Guidance for Quantifying and Using Long-Duration Truck Idling Emission Reductions in State Implementation Plans and Transportation Conformity*. EPA420-B-04-001. January 2004. http://www.epa.gov/smartway/idle-guid.htm

- U.S. EPA. *The Diesel Emissions Quantifier (DEQ).* This on-line interactive tool estimates the emissions reductions of clean diesel projects and their cost effectiveness. Accessible at: http://www.epa.gov/cleandiesel/
- U.S. EPA. *National Mobile Inventory Model.* This model estimates emissions from highway vehicles and nonroad equipment. Accessible at: http://www.epa.gov/otaq/nmim.htm
- U.S. EPA. *Smartway Technology Package Savings Calculator.* This calculator is designed to help truck owners compare the costs and estimate the fuel savings associated with various efficiency technologies. Accessible at: http://www.epa.gov/smartway/calculator/loancalc.htm
- U.S. EPA. *Guidance for Quantifying and Using Emission Reductions from Best Workplaces for Commuter Programs in State Implementation Plans and Transportation Conformity.* EPA420-B-05-016. October 2005. http://www.epa.gov/otaq/stateresources/policy/transp/commuter/420b07015.pdf
- U.S. EPA. *Diesel Retrofit Technology: An Analysis of the Cost Effectiveness of Reducing Particulate Matter Emissions from Heavy-Duty Diesel Engines Through Retrofits.* EPA420-S-06-002. March 2006. http://www.epa.gov/cleandiesel/publications.htm
- U.S. EPA. *Diesel Retrofit Technology: An Analysis of the Cost Effectiveness of Reducing Particulate Matter and Nitrogen Oxides Emissions from Heavy-Duty Nonroad Diesel Engines Through Retrofits.* EPA420-R-07-005. May 2007. http://www.epa.gov/clean diesel/publications.htm
- U.S. EPA. *Guidance for Quantifying and Using Long-Duration Switch Yard Locomotive Idling Emission Reductions in State Implementation Plans.* EPA420-B-04-002. January 2004. http://www.epa.gov/smartway/idle-guid.htm

EPA has prepared these documents and resources to help States and MPOs estimate emission benefits. The guidance documents address how to take credit for them in SIPs and conformity determinations.

These activities can be incorporated into the SIP as mandatory measures or as a voluntary mobile source measure as appropriate. Additional guidance documents and other information resources from EPA's Office of Transportation and Air Quality are available at: www.epa.gov/otaq

APPENDIX

Table 1. Summary of Cost-Effectiveness for Various Diesel PM Retrofit Scenarios

Vehicle	Retrofit Technology*	Range of $/ton PM Emission Reduced**	
School Bus	DOC	$12,000	$49,100
	CDPF	$12,400	$50,500
Class 6&7 Truck	DOC	$27,600	$67,900
	CDPF	$28,400	$69,900
Class 8b Truck	DOC	$11,100	$40,600
	CDPF	$12,100	$44,100
250 hp Bulldozer	DOC	$18,100	$49,700
	CDPF	n/a	n/a

* Retrofit technologies include diesel oxidation catalyst (DOC) and catalyzed diesel particulate filter (CDPF)

** The cost per ton of PM reduced will depend on a variety of factors including the age and activity levels of the vehicles or equipment. Source: U.S. EPA. *Diesel Retrofit Technology: An Analysis of the Cost Effectiveness of Reducing Particulate Matter Emissions from Heavy-Duty Diesel Engines Through Retrofits.* EPA420-S-06-002. March 2006. http://www.epa.gov/cleandiesel/publications.htm

Table 2. Summary of Cost-Effectiveness for Various Nonroad Diesel PM Retrofit Scenarios

Equipment	Retrofit Technology*	Range of $/ton of PM Reduced**	
Off-highway trucks	DOC	$21,700	$78,800
	CDPF	$24,200	$87,600
Loaders/Backhoes/Tractors	DOC	$25,900	$49,900
	CDPF	$28,800	$55,400
Excavators	DOC	$22,300	$61,900
	CDPF	$24,800	$68,800
Cranes	DOC	$20,900	$60,000
	CDPF	$23,300	$66,700
Generator Sets	DOC	$18,700	$46,100
	CDPF	$20,800	$51,300

* Retrofit technologies include diesel oxidation catalyst (DOC) and catalyzed diesel particulate filter (CDPF)

** The cost per ton of PM reduced will depend on a variety of factors including the age and activity levels of the vehicles or equipment. Source: U.S. EPA. *Diesel Retrofit Technology: An Analysis of the Cost Effectiveness of Reducing Particulate Matter and Nitrogen Oxides Emissions from Heavy-Duty Nonroad Diesel Engines Through Retrofits.* EPA420-R-07-005. May 2007. http://www.epa.gov/cleandiesel/publications.htm

Table 3. Summary Cost-Effectiveness of EPA Voluntary Mobile Source Programs (4:1 weighting of NO_x;VOC)

Activity	Description	Median Cost	Cost Range		Sources
Advanced Truck Stop Electrification	Advanced truck stop electrification provides a parked truck with electrical power, and heating, cooling, and other amenities like telecommunication hook ups, through an external console that fits into the truck's window frame.	$1,700	$1,400	$2,000	1
Truck Auxiliary Power Units	An APU consists of a small engine and generator that provides power to the truck when the main engine is shut off. It power s heating, air conditioning, and electrical accessories for the cab and sleeper.	$3,100	$2,700	$3,500	2
Diesel Retrofits	Examples of diesel retrofits include engine upgrades, engine repowers or replacements, cleaner fuels, emissions control technologies, or idle controls.	$5,950*	$1,900*	$19,000*	3
Best WorkPlaces for Commuters	Best Workplaces for Commuters[SM] is a voluntary business-government program that distinguishes and provides national recognition to employers offering outstanding commuter benefits such as free or low cost bus passes, strong telework programs, carpooling matching and vanpool subsidies. www.bwc.gov	$19,200	$19,200	$19,200	4

Sources: (1)NO_x emission factor based on EPA engine idling testing (see http://www.epa.gov/smartway/idle-guid.htm); assumes activity level of 40%-50% (or 9.6 -12 hours/day) utilization of one electrified parking space over 365 days/year (06/28/06 interview, Tom Badgett, Chief Information Officer, IdleAire Technologies, Inc.; assumes 20 year service life (06/26/06 interview, Tom Badgett); assumes $17K per space (see Texas A&M Research Foundation, Request for Proposals #B001498, http://rf-web.tamu.edu/); (2)Assumes 2,000 hour/year usage (interview with 3 leading APU manufacturers – Teleflex/Carrier, Rig Master, Thermo King; assumes NOx offset from APU at 23 g/hr (see http://www.epa.gov/smartway/idle-guid.htm; assumes10 year service life (interview with APU manufacturers); assumes $7K per unit (see American Transportation Research Inst. RFP #3-14-1104). (3) NOx reduction based on applying the methodology outlined in the technical paper "*Diesel Retrofit Technology: An Analysis of the Cost-Effectiveness of Reducing Particulate Matter Emissions from Heavy-Duty Diesel Engines Through Retrofits*" to a select group of NOx reducing retrofit technologies. The cost per ton of NOx reduced will depend on a variety of factors including the age and activity levels of the vehicles or equipment and the technology utilized. These cost figures only account for NOx reductions and do not factor in VOC reductions. Consequently, the 4:1 NOx:VOC weighting does not apply. (4) Best Workplaces for Commuters: E.H. Pechan and Associates, Inc., *PM NAAQS Mobile Modeling Technical Memorandum.* Prepared for U.S.EPA. July 2006.

Table 4. Cost-Effectiveness of Various Mobile Source Projects (4:1 weighting of NO_x:VOC)

Activity	Description	Median Cost (2006 dollars per ton)	Cost Range (2006 dollars per ton)	
Inspection and Maintenance	Typically pays for the operating expenses of I&M programs	$2,200	$2,100	$6,800
Regional Rideshare	Regional rideshare programs provide marketing, administrative and limited operating costs for area-wide carpool and ridesharing programs.	$8,700	$1,400	$18,700
Vanpool Programs	Similar to regional rideshare but focused on support and promotion of vanpools.	$12,300	$6,100	$104,200
Travel Demand Management	Usually informational and promotional programs administered by governments, public agencies and public-private partnerships. Often implemented at the employer level, utilizing ridesharing, transit and parking strategies	$14,600	$2,700	$38,900
Conventional Fuel Bus Replacement	Replace pre-1991 model year diesel buses with post-1996 diesel buses. Emission reductions and cost-effectiveness decrease rapidly if replacing newer buses	$18,800	$12,900	$46,700
Alternative Fuel Non-Transit Vehicles	Electric, CNG, LPG Vehicles and fueling facilities	$20,800	$4,700	$37,000
Traffic Signalization	Traffic signal inter-connection and timing optimization to reduce stop-and-go traffic.	$23,500	$7,000	$149,900
Employer Trip Reduction	Individual programs can vary widely. Can be voluntary or mandatory for employers but usually voluntary for employees. Employers offer workplace, work schedule, transportation and parking options along with incentives and disincentives for employee participation	$26,600	$6,800	$205,500
Transit - Conventional Service Upgrades	Consists largely of improved frequency of fixed-route bus service.	$28,800	$4,400	$140,600
Park & Ride Lots	Capital costs for construction of parking lots to support rideshare, express bus, HOV lanes	$50,300	$10,100	$82,800

Table 4. (Continued)

Activity	Description	Median Cost (2006 dollars per ton)	Cost Range (2006 dollars per ton)	
Modal Subsidies & Vouchers	Temporary subsidies to support new transit, carpool, or vanpool programs	$54,600	$900	$551,400
Transit - New Fixed Guideway and Equipment	Mostly fixed transit-ways and commuter rail capital investments. High initial capital costs, mediated by long service lives and projected long-term increases in ridership can produce favorable cost-effectiveness.	$77,700	$10,000	$551,200
Bicycle and Pedestrian Programs	Construction of new bicycle or pedestrian facilities, improved access to transit and activity centers, and education and safety program.	$98,500	$4,900	$403,600
Transit - Shuttles, Feeders, and Paratransit	Usually this is service and/or facilities to supplement regular public transportation routes. Shuttles and feeders are generally focused on specific corridors with identified demand for transit access. Paratransit serves wider areas with less demand - usually reflected in higher costs and lower impacts	$102,400	$14,400	$2,311,000
Alternative Fuel Transit Vehicles and Facilities	Predominately CNG replacement buses and shuttles	$148,000	$7,800	$665,800

Source: (1) Transportation Research Board. The Congestion Mitigation and Air Quality Improvement Program: Assessing Ten Years of Experience. Special Report 264. Washington, D.C.: 2002. In calculating cost-effectiveness, TRB assumed a weighting scheme of 1:4 for VOC:NOx. EPA has adjusted the TRB's calculations for inflation according to the consumer price index established by the Bureau of Labor Statistics, U.S. Department of Labor.

End Notes

[1] For a complete list of all EPA verified technologies, consult the list at the following web site: http://www.epa.gov/otaq/retrofit/retroverifiedlist.htm Information on CARB's verification program can be accessed at: http://www.arb.ca.gov/diesel/verdev/verdev.htm."

[2] 23 U.S.C. Section 149(f)(2)

[3] The Diesel Emissions Reduction program can be found in Subtitle G of the Energy Policy Act of 2005. Section 792(c)(3) specifies cost effectiveness as a priority for evaluating projects.

[4] Section 793(b)(1)(C)

[5] Transportation Research Board. *The Congestion Mitigation and Air Quality Improvement Program: Assessing 10 Years of Experience.* Special Report 264. Washington, D.C.: 2002.

[6] On October 31, 2006, the U.S. Department of Transportation issued interim guidance for the CMAQ Program as reauthorized under SAFETEA-LU. This document is available electronically at http://www.fhwa.dot.gov/environment/cmaq06gm.htm

[7] U.S. EPA (2002) Health Assessment Document for Diesel Engine Exhaust. EPA/600/8-90/057F Office of Research and Development, Washington DC. This document is available electronically at http://cfpub.epa.gov/ncea/cfm/recordisplay.cfm?deid=29060

[8] U.S.EPA (1996) Air Quality Criteria for Particulate Matter, EPA 600-P-95-001aF, EPA 600-P-95-001bF. This document is available in Docket EPA-HQ-OAR-2005-0036. U.S. EPA (2004) Air Quality Criteria for Particulate Matter (Oct 2004), Volume I Document No. EPA600/P-99/002aF and Volume II Document No. EPA600/P-99/002bF. This document is available in Docket EPA-HQ-OAR-2005-0036. U.S. EPA (2005) Review of the National Ambient Air Quality Standard for Particulate Matter: Policy Assessment of Scientific and Technical Information, OAQPS Staff Paper. EPA-452/R-05-005. This document is available in Docket EPA-HQ-OAR-2005-0036.

[9] U.S. EPA. Air Quality Criteria for Ozone and Related Photochemical Oxidants (Final). U.S. Environmental Protection Agency, Washington, DC, EPA/600/R-05/004aF-cF, 2006. This document is available at: http://www.epa.gov/ttn/naaqs/standards/ozone/s_o3_cr_cd.html

[10] For a complete list of all EPA verified technologies, consult the list at the following web site: http://www.epa.gov/otaq/retrofit/retroverifiedlist.htm. The MOA can also be found at this web site.

[11] A list of idle-control technologies can be found at: http://www.epa.gov/otaq/smartway/idling technologies.htm

[12] For more information, see: http://www.epa.gov/otaq/models.htm and http://www.epa.gov/omswww/stateresources/tools.htm EPA has also recently released a quantification tool for estimating environmental impacts and cost effectiveness of emissions reduction technologies added to vehicles and equipment. The Diesel Emissions Quantifier can be accessed at: http://cfpub.epa.gov/quantifier/

[13] See the EPA Clearinghouse for Inventories and Emissions Factors - 2002 National Emissions Inventory Data and Documentation. http://www.epa.gov/ttn/chief/net/2002inventory.html

[14] Because the current $PM_{2.5}$ designations were effective in 2005, areas must demonstrate attainment as expeditiously as practical but no later than 2010. However, Clean Air Act section 172 (a)(2)(A) allows areas to request that the attainment date be extended by up to five additional years if they can justify the need for such an extension based on the severity of the nonattainment problem and the availability and feasibility of control measures.

[15] Clean Air Act section 107(d)(1)(B) requires EPA to promulgate designations within two years of the promulgation or revision of a NAAQS. However, this time period may be extended for up to one year if there is insufficient information to promulgate the designations.

INDEX

C

D

E

F

M

N

O

P

Q

R

S

T

U

V

W

Y